# SYSTEMIC THINKING

# SYSTEMIC THINKING
## Building Maps for
## Worlds of Systems

**JOHN BOARDMAN**
John Boardman Associates
Worcester, UK

**BRIAN SAUSER**
University of North Texas
Denton, TX

For a copy of SystemiTool see http://www.WorldsofSystems.com.

For general information on our other products and services or for technical
support, please contact our Customer Care Department within the United States
at (800) 762-2974, outside the United States at (317) 572-3993 or fax
(317) 572-4002.

Wiley also publishes its books in a variety of electronic formats. Some content
that appears in print may not be available in electronic formats. For more
information about Wiley products, visit our web site at www.wiley.com.

*Library of Congress Cataloging-in-Publication Data is available.*
ISBN 978-1-118-37646-1

Printed in the United States of America

10  9  8  7  6  5  4  3  2  1

# CONTENTS

# LIST OF SYSTEMIGRAMS

# LIST OF FIGURES

# LIST OF TABLES

# ACKNOWLEDGMENTS

Don't walk behind me; I may not lead. Don't walk in front of me;
I may not follow. Just walk beside me and be my friend.

—Albert Camus

Once riding a train from Hobokcn to Aberdeen, New Jersey, Brian
turned to John and said, "We need to write the 'how-to' guide for
systemigrams." What transpired the rest of that train ride and later
at Brian's house over a few glasses of scotch was what you are
about to read. This book is the evolution of an idea born of passion,
built on a legacy seeded by systems pioneers, and the work of two
people inspired by others and the powers above them. If you ask
us who wrote this book, we will tell you, "We do not know." It is a
book written by many as we have walked through the worlds of
systems. As for our journey for this book, we have been influenced
by many, but can mention only a few. These people have challenged
our ideas, inspired us to new dimensions, and helped us open our
box so we could step outside of it. They are Clif Baldwin, Jennifer
Bayuk, Donny Blair, Alison Boardman, Robert Cloutier, Robert
Edson, John Farr, Ralph Giffin, Alex Gorod, Eirik Hole, Larry
John, George Korfiatis, Ryan McCullough, David Nowicki, Michael

Pennotti, Wesley Randall, Andy Taylor, Dinesh Verma, and Jon Wade.

### From John
A system is relatively simple to define, but never easy to build. Parts don't fit, relationships break, and wholes become dysfunctional and disintegrate. If my partnership with Brian Sauser is considered a system, it is exceptional. And that speaks volumes for Brian, and explains why this volume will speak compellingly to you. Brian, may our system endure. I'm hoping forever.

My darling wife Alison, how on earth do you put up with me? It is because heaven is in your heart and so it becomes mine and ours. Together, we rejoice that our system is blissfully enhanced with the arrival and growth of each grandchild. May they learn how to be different and to fit in; may they discover relationships and contribute fully to them all; may they bring wholeness and so be channels of blessing to many worlds of systems.

### From Brian
In our book *Systems Thinking: Coping with 21st Century Problems*, I wrote, "I stand at the edge of the future with the universe as my systems boundary and standing beside me is what I believe will be the greatest of my systems shepherds, John Boardman." John, my sentiment has never wavered, and I will never leave the flock without the light of your candle to show me the way.

My wife, who loves me not because it is right or necessary, but because it is. It is love; it is commitment; it is sacrifice; it is unconditional; it is. To my children, who will forever try to figure out what Daddy does for a living, I ask you to remember one thing, and that is passion. Passion is what wakes you up in the morning; passion is why you find meaning in tomorrow; passion fills the cracks in the sidewalk of life. Attack whatever you do with passion, and you will find solace in everything that you do.

# JOURNEY I

## SYSTEMIC FAILURE

# CHAPTER 1

# WHERE WE START FROM

What we call the beginning is often the end
And to make an end is to make a beginning.
The end is where we start from.

—T.S. Eliot, *Four Quartets*

This is a book about problem-solving, but with a difference. We recognize three vital characteristics, which for far too long have been overlooked or neglected by problem-solving books.

First, we identify that while solutions undoubtedly "deal with" the problems to which they relate, they also create a new wave of problems in their wake. In our complex world, this problem-generating characteristic of solutions cannot be ignored, and problem-solving itself must take care not to become problem-spreading in nature. It has been widely recognized for some time that problems themselves can spread or cascade, as in the case of electricity

*Systemic Thinking: Building Maps for Worlds of Systems*, First Edition.
John Boardman and Brian Sauser.
© 2013 John Wiley & Sons, Inc., Published 2013 by John Wiley & Sons, Inc.

supply networks (e.g. New York City blackout) or the growth of cancer in the human body (e.g. prostate cancer in adult males). But the realization of problems elsewhere caused by the creation of a solution in some particular area of interest, removed from these affected other regions, is both alarming and unsettling. The way forward that we propose in this book gives due recognition to this phenomenon.

Second, the emergence of a class of person known as problem-solver, identified by skills in problem-solving, has reduced the burden on the class known as problem-owner, to the extent that the latter has effectively transferred the problem and subsequently lost ownership, and in so doing created a false picture for the former who cannot therefore avoid endowing the solution with the problem-spreading gene. This distinction of classes, one which effectively divorces the two, must be overcome, and problem-solving in our complex world must restore the vitality of problem-ownership among those who sense the problem in the first instance.

The third characteristic is something we can more easily recognize if we stand back from the first two. When a solution to a given problem also leads to a wave of new problems, then problem-solving essentially becomes problem-spreading. When problem-solving attracts a new breed of person, who become known as problem-solvers, then responsibility for the problem is in effect transferred—from those it first affects or who sense it, with attendant diminution in problem-ownership. We might say problem-solving becomes problem-dispossession. So standing back leads us to conclude that the originating problem is strongly connected to a host of "accompanying apparatus," including owners, solvers, and problem-solving approaches. It is this connectedness that marks out this third characteristic, which we believe has hitherto been sorely neglected and about which this book has much to say. Moreover, this book has much to offer by way of a responsive way forward.

Our way forward is what we call *systemic thinking*. It is a way of thinking that emphasizes connectedness and enables people to see the bigger picture; one in which owners, solvers, solutions, problem-solving methods, and problem descriptions are portrayed as a whole system.

As you traverse through this book, we see it as a passage into Worlds of Systems. As such, the book is in three parts, which we have rightfully named Journeys. We sincerely hope that these Journeys will form a coherent whole, that when you are done will bring you to a place you were not before you started. In Journey I, we describe systemic failure—an increasingly popular term among politicians *inter alia* for describing the meltdowns and near catastrophes involving multiple stakeholders and systems—as the representation of problems, which cascade. This term applies when there is evident lack of problem ownership coupled with piece-meal approach to problem-solving and reliance on unsustainable solutions.

When confronted with a problem that appears to be without solution, we apply frameworks from our intellect to shine a light on a potential path. In Journey II, we present a system of ideas, which helps us to form a language that better enables us to describe specific systemic failures, and in so doing forming more well-rounded problem descriptions. This is our framework for enlightening a path.

In Journey III, we introduce the idea of systemic diagrams, which we call systemigrams. These are our maps to systemic problems. We provide numerous examples of specific instances of how systemigrams have helped overcome piecemeal problem-solving by drawing together owners, solvers, problem descriptions, and relevant solution. Journey III gives the reader a comprehensive opportunity to learn what systemigrams are, how they are created and put to effective use, and why they are an efficacious approach to complex problem-solving.

These are our journeys into Worlds of Systems and systemic thinking.

# CHAPTER 2

# SYSTEMIC INTRODUCTION

On Christmas Day 2009, 19-year-old Farouk Abdulmutallab and a few highly dedicated cohorts exposed significant flaws in an extended enterprise comprising at least "1271 government organizations and 1931 private companies" and a combined annual budget in excess of $75 billion (Priest and Arkin 2010). The reason Abdulmutallab's operation did not rain death and destruction down on the people of Detroit, Michigan, was that when the terrorist unsuccessfully attempted to ignite his explosives package, an alert Dutch passenger took action to subdue him. Consequently, the passengers on that flight, the people of Detroit, and the people of the United States of America got very lucky.

The Senate Select Committee on Intelligence (SSCI) was not as generous in its conclusions as was the newly installed Secretary of Homeland Security, Janet Napolitano. According to the Executive Summary of the SSCI's May 2010 report on the subject (SSCI 2010, May 18), the U.S. Counterterrorism Enterprise failed to do the job

*Systemic Thinking: Building Maps for Worlds of Systems*, First Edition.
John Boardman and Brian Sauser.
© 2013 John Wiley & Sons, Inc., Published 2013 by John Wiley & Sons, Inc.

for which it was created, not because of any compelling flaw in its many technological capabilities, but because its members chose not to share critical information in their possession—they chose to not cooperate with each other.

President Obama broke off his holiday in Hawaii for a second time in as many days to address the nation over Farouk Abdulmutallab's failed attack on Northwest Airlines 253, which flew from Amsterdam headed for Detroit. Obama delivered a blunt admission that the system designed to protect Americans in the wake of the 9/11 attacks had failed, calling the breakdown "totally unacceptable." In a December 29, 2009, press conference in Kaneohe, Hawaii, the President said (Obama 2009):

> When our government has information on a known extremist, and that information is not shared and acted upon as it should have been, so that this extremist boards a plane with dangerous explosives that could cost nearly 300 lives, a systemic failure has occurred and I consider that totally unacceptable. There was a mix of human and systemic failures that contributed to this potentially catastrophic breach of security. We need to learn from this episode and act quickly to fix the flaws in our system because our security is at stake and lives are at stake.

"Systemic failure." Very interesting. What is it?

In a speech at the Council on Foreign Relations in Washington, DC made on March 10, 2009, Federal Reserve Chairman Ben Bernanke used the term systemic (or systemically) a total of 37 times. Clearly, he was not referring to events yet to be some 9 months hence. Instead, his topic of interest was financial reform, which gained in increasing urgency as the global economy continued to struggle with the aftermath of bank and insurance company collapses in the wake of Lehman Brothers, the United States's fourth largest bank, filing for bankruptcy on September 15, 2008.

In his speech, Bernanke attributes the origins of global economic meltdown "to the global imbalances in trade and capital flows that began in the latter half of the 1990s." Basically, burgeoning Asian economies coupled with prudent savings from an emerging middle class produced huge mountains of capital, eagerly desired by Western economies, with voracious consumer appetites, little interest in saving, and possessed of an inordinate confidence in spending on assets that were believed could only increase in value, for example, property and homes. Bernanke remarked,

> Like water seeking its level, saving flowed from where it was abundant to where it was deficient, with the result that the United States and some other advanced countries experienced large capital inflows for more than a decade, even as real long-term interest rates remained low.

The Fed Chairman continued, "the risk-management systems of the private sector and government oversight of the financial sector in the United States . . . failed to ensure that the inrush of capital was prudently invested, a failure that has led to a powerful reversal in investor sentiment and a seizing up of credit markets. When those failures became evident, investors lost confidence and crises ensued."

Reform was both essential and inevitable, and in Bernanke's thinking this meant: "We must have a strategy that regulates the financial system as a whole, in a holistic way, not just its individual components. In particular, strong and effective regulation and supervision of banking institutions, although necessary for reducing systemic risk, are not sufficient by themselves to achieve this aim." He adds, "We should consider whether the creation of an authority specifically charged with monitoring and addressing systemic risks would help protect the system from financial crises like the one we are currently experiencing."

These terms—"systemic risk," "systemic failure," and allow us to include "systemic remedy"—are they just clever phrases used by smart people to confound ordinary citizens and so devalue common sense? Or might they point to something quite fundamental that even the intelligentsia have missed, something that calls for a reformation in all our thinking?

An interesting question we would like to pose is, "How can such diverse situations (national security and international finance) and vastly differing circumstances be given the same label, that of systemic failure?" Clearly, both were failures, but more than that, as completely different as they are, they can be identically described as "systemic failures." Why? And how? What exactly is systemic failure?

We know that systems sometimes fail. The autopilot in a modern airliner can fail. A set of traffic lights at a busy highway intersection can fail. Your liver can fail. Each one of these, autopilot, traffic lights, and human liver can be regarded as systems in their own right, of varying degrees of composition and complication, and when any one of them fails it can be called a system failure. A system failure but not a systemic failure. What's the difference? To answer that question we have to see these vastly differing objects not only as systems but also as parts.

The failed autopilot can lead to the crash of an airliner and the subsequent crash of an airline. The reality is that systems do not exist in isolation. They live as parts in a greater system by virtue of myriad connections not all of which are obvious many in fact being very subtle, as exemplified by the interconnections between sociological and technological systems in the case of the traffic lights.

The failure of a set of traffic lights at a busy intersection can cause grid lock, road rage, civil disobedience, missed appointments, lost business opportunities, and if not the death of road users then conceivably the death of commercial contracts and the demise of corporations. It's very unlikely, but it is possible. Once again, because systems are interconnected, forming a greater system, a sequence of cascading failures leads to total system meltdown. That's systemic failure.

If your liver fails, there are consequences, and this is because the liver plays an important part in your body. The liver is indeed a system but it is also a vital part of a greater system because of the physiology of the human body.

The consequential effects of liver failure, depending on the extent of that system failure, can lead to damage and subsequent failure of other connected parts, indeed other systems, setting up a set of cascading system failures that can eventually lead to a total system shutdown. Death. That's systemic failure.

Systems are also parts, and as such they inhabit worlds of systems, though not always as prominent or even self-evident members.

The truth is that whereas systems are very familiar to us, these worlds of systems are largely unknown to us and we discover their existence all too tragically via systemic failure.

Is it possible, we ask, for these worlds to be adventured, explored, mapped, and navigated in advance so that when system failures do occur, systemic failure itself can be avoided? That's what this book is all about—building maps of these worlds of systems using systemic thinking as the natural antidote of systemic failure. And if you want to know more about the nature of systemic failure, then we hope this book will help you achieve mastery over systemic thinking, which will then help you to play your part, as a complex system, in anticipating and avoiding systemic failure and so prevent it from being a menace to you and to many others.

System failures we can live with—usually. Systemic failure, however, poses far greater challenges. The financial crisis in recent years almost led to the total collapse of the world's economic systems. No one knew what the consequences of that would have been. It didn't happen and yet it might still. That is risky, and it is troubling that people cannot agree on what to do about it. Troubling but understandable. Because the root of systemic failure lies in our ignorance of these worlds of systems in which market forces, government regulations, and human desires for a continually improving quality of life interact via complex relationships.

We are not going to stop building systems. They are an essential commodity to all of us. What we need to do is understand that the boundary of a system is not the end of it. More likely, but less obviously, it's the beginning of it. Because that system, once it has begun life, becomes a member of worlds of systems. So what we must do in order to avoid systemic failure is to *understand systems as parts*, and to do that we need systemic thinking.

Professionals of all walks of life have benefited from systems thinking, the bodies of knowledge they use in order to create systems of all kinds, for a very long period of time. The interconnected age we live in now demands a new form of thinking that deals with systems as parts and that calls for systemic thinking, a huge part of which is the skill to build maps of these worlds of systems.

# CHAPTER 3

# RAINING ON MY CASCADE

Buchanan, New York, is conspicuous for its anonymity. Except of course to the village's residents, who number around 2000. To them, Buchanan is home. For the millions in New York City, the location is unknown and of no significance. The Big Apple's teeming masses go about their lives blissfully unaware of Buchanan's existence and probably of the fact that the nuclear power plant of Indian Point lies on its doorstep. Unaware and largely uncaring, that is, until their lights go out with consequences that if not considered seismic, then certainly are unignorable.

In the stifling heat of New York summers, air conditioners run perpetually and obediently, keeping homes and offices at agreeable room temperatures for their occupants to rest and work. Meanwhile, baseball fans cheer energetically for their team as players slug it out for victory under floodlit skies. It is life and business as usual, for Americans and for the attendant weather systems that predictably are equally energetic at that time of year.

*Systemic Thinking: Building Maps for Worlds of Systems*, First Edition.
John Boardman and Brian Sauser.
© 2013 John Wiley & Sons, Inc., Published 2013 by John Wiley & Sons, Inc.

On the evening of Wednesday, July 13, 1977, a lightning strike at Buchanan South substation tripped two circuit breakers. A loose locking nut combined with a tardy upgrade cycle ensured that the breaker was not able to reclose and allow power to flow again. A second lightning strike caused the loss of two 345 kV transmission lines, subsequent reclose of only one of the lines, and the loss of power from the nearby 900 MW nuclear plant. As a result of the strikes, two other major transmission lines became overloaded beyond normal limits. As per procedure, Consolidated Edison (Con Ed), the power provider for New York City and some of Westchester County, tried to fast-start generation at 8:45 p.m. EDT; however, no one was manning the station, and the remote start failed. After this second failure, Con Ed had to manually reduce the loading on another local generator at their East River facility, due to problems at the plant. This exacerbated an already dire situation.

If load, which is to say the demand for electricity from consumers, remains constant and yet the routes of delivering power, which itself is in plentiful supply, begin to close, either from outages due to lightning strikes or operator-induced open circuits to avoid damage to equipment, then the remaining healthy delivery routes come under exceeding stress. Inevitably, these too must be closed off. It is an accelerating phenomenon that is commonly described as cascading failure. The consequences are that demand is not met even though there is supply at source, and the lights go out, which is to say the power is cut off.

The consequences of the 1977 blackout included the following: LaGuardia and Kennedy airports were closed for about 8 hours; automobile tunnels were closed because of lack of ventilation; 4000 people were evacuated from the subway system; Shea Stadium went dark while the Mets were losing to the Chicago Cubs; most of the city's television stations were taken off air; and looting and vandalism became prevalent, with some veterans of the 1965 blackout taking to the streets at the first sign of darkness. In this latter case, the temptation to illegally acquire couches, televisions, and heaps of clothing from neighborhood stores was irresistible to an increasing number given the darkness, the lack of policing with officers themselves hampered by the blackout, and the readymade excuse of a prevailing economic depression.

Before the lights came back on, even Brooks Brothers on Madison Avenue was looted. In all, almost 2000 stores were damaged in looting and rioting; more than 1000 fires were responded to, including 14 multiple-alarm fires; and, in the largest mass arrest in city history, nearly 4000 people were arrested. Many had to be stuffed into overcrowded cells, precinct basements, and other makeshift holding pens.

A congressional study estimated that the cost of damages amounted to a little over US$300 million. The city was later given over $11 million dollars by President Carter's administration to pay for the damages of the blackout.

Con Ed called the shutdown an "act of God," enraging the city's Mayor Abe Beame, who charged that the utility was guilty of "gross negligence." Beame himself later suffered in the aftermath, coming in third in the Democratic primary later that year, a race won by Ed Koch, who went on to become the city's new mayor.

How can a couple of lightning strikes in the middle of nowhere bring the largest city in the United States virtually to its knees and cause its mayor's head to roll even as the city's poor help themselves to serendipitous bounty? How? Cascading failure, that's how: a synonym for systemic failure.

It's not that it's nobody's fault. But it is that it is no *one* fault. Rather, it is a series of cascading failures where the interconnections on many levels make each failure a contributor to the next. This series or sequence occurs because there exists an unknown world of systems that emerges as we continue to build individual systems. We little realize that these systems form their own constituency and populate a world unknown to us.

A circuit breaker is truly a marvel of engineering. It is an automatically operated electrical switch designed to protect an electrical circuit from damage caused by overload or short-circuit. Its basic function is to detect a fault condition and, by interrupting the continuity of supply, to immediately discontinue electrical flow. Whereas a fuse operates once in cutting off supply in an electrical circuit that would otherwise overload, and then has to be replaced, a circuit breaker can be reset (either manually or automatically) to resume normal operation.

Circuit breakers are made in varying sizes, from small devices that protect an individual household appliance up to large switchgear designed to protect high-voltage circuits feeding an entire city.

The automatic operation of the circuit breaker means that it exists both to protect and to serve (not unlike a police force). It protects electrical apparatus from the shock of excessive power flow and, once the shock has passed or been dealt with, it can restore power flow and continue to serve the demands for electricity consumption.

It would be highly desirable to invent and "engineer" circuit breakers that protect and serve systems other than the electrical variety, namely, those systems that fundamentally depend on the use of electricity: in other words, society itself and communities within society, for example, users of the subway and the automobile tunnels, shopkeepers vulnerable to impromptu marauders, and city mayors who carry the can come election time for matters over which they had little control. Such an invention, *a kind of "social circuit breaker,"* would indeed be marvelous, but at present its design appears to present an intractable problem.

When a circuit breaker trips, the electrical system that is being protected is effectively shut down, and its users are left stranded. This does not immediately threaten their well-being, necessarily. And most of us realize that we will face such interruptions of service from time to time, and we learn how to cope with being cut off, for a short period of time. The circuit breaker "knows" this and works hard to restore supply and bring users back into their familiar routines as quickly and as safely as practicable. Normal service is restored as soon as possible.

The smarter these circuit breakers can become, relative to the consumers of electricity and not merely the electrical apparatus itself, the better. In other words, the connections between technology and society, between the electrical system and the social system for whom it has been provided, are the focus of our interest. It is this particular focus that drives so much of technology providers' efforts to build smarter systems, for example, cities and the infrastructures upon which these rely.

A greater challenge, however, is posed by the need to design and build "circuit breakers" within the social systems. For example, we might ask, "What linkages can be provided that automatically open and later reclose between individuals and small groups of people that will reasonably allow a herding instinct and yet prevent a stampede?" Examples of this would be a crowd's rush for safety from a perceived danger or a dash to possess the goods of others as a looting and vandalism mindset takes hold within a community. We can of course intend "stampede" to be taken both literally and metaphorically.

It is the susceptibility to influence of individual people that sets up linkages that under certain circumstances turn an innocent gathering into an unruly mob. Within each of us, though, lies the opportunity to act as an automatic circuit breaker, isolating shocks so that they don't ripple through the system and restoring common sense in the flow of responsible communications.

Systemic risk is posed and systemic failure occurs in worlds of systems in which linkages between systems are themselves faulty because they have not been purposefully designed as such but have emerged by virtue of an increasing population of systems that take it upon themselves to form their own connections.

That is bad enough, but worse is the reality that many of these interconnections go entirely unobserved and come into our consciousness only after systemic failure occurs. To rectify this condition requires us to know what we do not know and resolve to explore this unknown.

Understanding these linkages first and protecting them once identified goes a long way to reducing risk and avoiding failure, but to achieve this understanding requires an exploration of these largely unchartered territories, these mysterious worlds of systems.

# CHAPTER 4

# IT'S THE WHOLE, STUPID!

Why should baseball fans in Shea Stadium have their enjoyment of the game interrupted simply because a recondite piece of equipment nowhere remotely near Queens gets hit by a random lightning strike? It makes no sense. It's simply not fair: an event having nothing whatsoever to do with these fans being thrilled that causes a dumb piece of electrical apparatus to get overexcited while extinguishing their joy. Explain that!

The simple answer to this perfectly reasonable query is that Shea Stadium, like almost every other consumer of electricity, is dependent on electrical power that is first generated at one of many very remote locations, often in "the middle of nowhere," because nobody wants them in his or her backyard, and consequently that power must be delivered, in near real time, via an elaborate electricity distribution network and not simply in tidy discrete packages directly from point of origin to point of consumption. The electric-

*Systemic Thinking: Building Maps for Worlds of Systems*, First Edition.
John Boardman and Brian Sauser.
© 2013 John Wiley & Sons, Inc., Published 2013 by John Wiley & Sons, Inc.

ity for Shea is part of a huge continuum that also supplies the streets of midtown Manhattan and the suburbs of Queens.

Now if there's one thing you can say about a network, it's that it is a system. The network is made up of many parts, including substations, switchgear, circuit breakers, and the like, and many connections, such as overhead power lines. These parts and connections then combine to form a whole that we can rightly refer to as a system. Some engineers understand the parts, very thoroughly, and a few understand the system as a whole. The latter comprehension is achieved by having a variety of representations or models of this whole, and also by means of simulations, such as power flow analyses, which will include fault flows that tell the engineer what the power flows in the network will look like given a variety of fault conditions, for example, when a circuit breaker opens due to a lightning strike.

It is a professional and ethical responsibility on the part of the engineering community that looks after the electrical supply network to understand what can go wrong with and within their system for a wide variety of credible scenarios. If for no other reason, this is why situations like the 1977 New York City blackout do not last forever and in fact are overcome in far less time than would be the case without such an understanding of the system. Interruptions to the service, engineers will tell us, are bound to happen at some point in time. They reassure us that normal service is being resumed with all possible safety and speed. Fans at the ballgame may not be that impressed. Voters on election day show their disapproval. Circuit breakers are not the only thing to be struck down, as we have seen.

It is comforting to know that there are those who do understand the system and can make it work again, even though parts of it are taken out of action. If only that were true of other things in life, for example, the financial industry, which had its own shock circa 2008. The most alarming thing about this near-meltdown is that there were no systems engineers or systems professionals of any kind who actually knew what this system looked like. There were no network diagrams, no models, and no simulations. A few knew a lot about the parts of this system, like a bank or a credit card company, or a retail franchise, or an automobile loan company. These elements were well understood and more or less properly managed. But that is precisely what they were: elements, systems in their own right and yet parts of a greater whole. Pieces of a larger puzzle as it turned out, because, apparently, the whole, the financial system itself was little understood by an elite few at best, and at worse tragically unknown to anyone. It was and is in fact a world of systems with hardly any maps and little or no epistemology with which it could be analyzed and managed. That is why it's not out of character to use terms such as "systemic risk" and "systemic failure" and latterly "systemic oversight" relative to this particular large and complex system.

But before we turn to this sophisticated language, we should once again demonstrate our esteem of common sense by reference to Figure 4.1.

Who would be dumb enough to sit in a boat that is sinking and give thanks for the fact that the hole wasn't at his end of the boat? We don't know, but apparently some dumb people are capable of picking up seven-figure bonuses. Rather than envy the well-heeled investment banker or the slick derivative salesman or whomever, we should be rebuking them with the line, "It's not the hole, it's the whole, stupid!"

**Figure 4.1.** Sinking Boat

The Lehman Brothers' boat sank. This was not the only one. A veritable flotilla of ocean-going financial services corporations also sank. Drowned in waters far too deep and ludicrously uncharted for them to be floating on in the first place. Some boats were designated too big to drown. And so the USS Treasury was dispatched to issue lifeboats, life vests, and repair vessels to patch up the holes. Finally, they got to grips with the right whole. The model in most people's heads was that there was a flotilla. It was every man for himself, women and children, too. Each had his or her own boat. If one went down, too bad. It happens. Maybe more than one, maybe a few, maybe a large number go down. But not all. The big boats don't sink. Except the model was wrong. There really is only one boat, and "it's the economy, stupid!"

The real economy, that dimension of human society that has ideas, turns them into products and services, trades these in the marketplace, hires people, makes profit, stimulates others to trade, pays wages and dividends, makes donations to charity—this is the boat. And it includes the financial services industry because the real economy cannot operate without money, credit, loans, and banking services at large.

The real boat, the one true boat, is a world of these systems: of banks and businesses, of customers and credit, of leases and lenders, of interest and investors. And that boat, while it can stand some shocks, and pretty big ones at that, has a vulnerability, an Achilles' heel, just like any other boat. One that can sink it. What we don't need is someone in the boat who thinks that because the hole is at the other end, that person is safe. What we do need, as best as we can make it, is a picture of this boat, a map of this floating world that shows us where vulnerabilities lie. A map that shows us how these "small" systems are hooked together, and what can happen—in short order—if these small systems turn into holes endangering the whole real economy boat. That's systemic thinking.

In that marvelous movie *Up in the Air*, an intrepid novice fresh out of college offers her employees a revolutionary idea. The company she works for is a firm of consultants who are hired by other companies to do their dirty work for them, namely, to fire employees in the name of downsizing. Because this work involves the personal touch, with consultants clocking up millions of air miles in order to deliver the bad news to those being let go in face-to-face meetings, it's an expensive operation for the firm. The preppy's idea is to combine the global reach of the firm with the local touch using clever information technology. She describes her idea of *glocal*, a rather unsubtle portmanteau word formed from global and local.

Faced with the financial crisis of 2007 and onward, the U.S. government had to identify its own radical *glocal* approach, one that could look after the entire financial services industry, rescuing it from the abyss and safeguarding its future stability, by appropriate supervision and regulation of its constituent bank holding companies (BHC). The industry in its entirety is the global piece of the puzzle, while the BHCs, latterly including Goldman Sachs, Morgan Stanley, American Express, and General Motors Acceptance Corporation (GMAC), are the local elements. The government wanted to take care of the flotilla and each boat, recognizing in the nick of time that a boat failure could capsize the whole flotilla. To put rigor into their approach, it was first necessary to identify how a local failure could lead to systemic risk and conceivably the collapse of the entire system.

It is unusual for a bank to fail. Unusual but not unheard-of. Why might it happen and what are the consequences? The fractional reserve system enables a bank to retain only a percentage of its capital, monies provided by depositors. The balance is lent out, maybe to some of those self-same depositors, for the purposes of credit and for what borrowers regard as attractive investments, whereby they can service (and ultimately repay) the debt and in the longer term see a reward for that investment. Borrowers take a risk. But it's with the bank's money. A loan is actually regarded by the bank as an asset, and amazingly, this asset can also be loaned out, thanks to the fractional reserve system. If the bank is wise, it will ensure the creditworthiness of borrowers. However, when the incentives to loan are high, such vigilance may attenuate. For example, if the bank can borrow money cheaply and lend it out at much higher rates, making money for itself immediately via loan arrangement fees, it is extraordinarily tempting for a bank to do this business. This is particularly the case if banks can hedge their risks by insuring these loans against default with other banks or insurance companies that also make up the flotilla. It does not take much imagination to see how links between these autonomous financial systems can quickly grow and in the process become invisible. Seeing the big picture becomes a challenge especially when the need to do so is obscured by the latent confidence we all appear to have in each element and in the notion that failures can be quickly isolated, much like faults on an electrical network can be isolated, preventing danger elsewhere in the network. However, financial "blackouts" are still possible.

Ensuring the well-being of any single BHC is termed "microprudential." This seeks to insure depositors against loss (up to $250,000). That might seem enough for most of us. But it could be a drop in the ocean if you're another bank, which has taken on some of the risk that eventually occurs when debtors default and there's little prospect of recovering the loss above a few cents on the dollar. However, it is easy to see how the pain of loss does not go away. On the contrary, it can spread. If a bank goes under, this puts stress on depositors who have lost their money. Others lose confidence in their own bank, especially if they suspect that the bank is a creditor of the expired bank. Word spreads and with it the pain for surviving banks. When bank depositors are themselves stressed, this has an effect on the economy via loss of consumer confidence and a diminution in corporate optimism. Spending and lending are reined in, further depressing the economy. The systems professionals call this positive feedback, which accounts for both healthy growth—or a virtuous cycle—as well as retrenchment—or a vicious cycle. It is as if a sinking boat sets up turbulence in the waters, destabilizing neighbors who may then also sink, thereby further amplifying these hostile waves. Before anyone can do much about it, a tsunami is in the making, one from which no boat is safe, where none is too big to fail. Hence the need for a macroprudential approach to be combined with the microprudential approach. For the U.S. government, this has appeared as the Supervisory Capital Assessment Program (SCAP).

It is not easy for us mere mortals to decode the language of the financial services industry, which seems incapable of demotic argot. But that is what we now seek to do using as far as possible terms and ideas that will serve our need for a cartography of these worlds of systems that simply emerge from myriad links of more elemental autonomous systems. SCAP is colloquially known as a bank stress test, and in the simplest possible terms, it is a way of ensuring that a BHC has enough capital to withstand the worst scenario imaginable. If it does not, then in order for it to continue, it must raise that capital from new investors or, as a last line of resort, from the Federal Reserve. As a consequence, the normal capital levels of the BHC, in keeping with the fractional reserve system, are enhanced by a SCAP buffer ensuring the outcome will withstand the more adverse scenario in possible future outcomes recognizing the systemic nature of the industry. If the more benign scenario obtains in the future, the BHC enjoys large capital surpluses.

An interesting comparison has been made between the microprudential, the macroprudential, and the SCAP approaches, summarized in terms of objective, impact, focus, risk exposure, perspective, and disclosure. SCAP has emphasized the linkages between BHCs and the real economy. It is its own crude map of how the boats make up the flotilla, at least in terms of systemic risk and prevention. SCAP's objective is to ensure adequate system capital to promote lending and restore consumer confidence and is explicitly designed to reduce the probability of adverse outcomes. Its focus was to examine the 19 largest BHCs possessing two-thirds of system assets among them. Risk exposure is estimated by applying common shocks to all participating BHCs incorporating idiosyncratic exposures and variations. It is the financial system equivalent to the electrical engineers' power and fault flow simulations. SCAP takes a 2-year perspective of potential performance in a low probability scenario and, in terms of disclosure, releases BHC-specific information about potential losses, resources, and capital needs. This aspiration after transparency is admirable. After all, what use is a map if it's not made available to travelers or, in this case, to the users of the financial system?

# CHAPTER 5

# THE ANSWER IS . . . PIONEER ACORN PANCAKES?

There is a story that has been written and told many times over about "whole systems thinking." It has been most commonly referred to as "Operation Cat Drop." It is best detailed by Patrick O'Shaughnessy, a professor at the University of Iowa (http://www. catdrop.com), but here is one such telling of the story that appeared in the December 18, 2005, edition of *The Star-Ledger* by Amy Ellis Nutt:

### Cats and a classic misstep

At the heart of social network analysis is "whole systems thinking"—which posits the view that when a decision is made in one place, there may be unforseen repercussions.

The classic example of failure to use whole systems thinking occurred in the 1950s when the World Health Organization tried to help a tribe on the island of Borneo. A mosquito and housefly infestation

*Systemic Thinking: Building Maps for Worlds of Systems*, First Edition.
John Boardman and Brian Sauser.
© 2013 John Wiley & Sons, Inc., Published 2013 by John Wiley & Sons, Inc.

had caused an outbreak of malaria and WHO sent planes to spray the island with DDT. (The pesticide was legal at the time.)

The DDT killed off mosquitoes, but also a species of wasp that preyed on thatch-eating caterpillars. With no wasps to dine on them, the caterpillars munched away on the thatched roofs of the tribes people's homes until the roofs caved in.

Worse, the poisoned houseflies were eaten by the island's geckos and the sickened geckos became easy eating for the cats. When the cats died from the accumulated DDT, the rat population flourished.

The result: outbreaks of typhus and plague. Faced with a health situation more dire than the original problem, WHO set in motion "Operation Cat Drop," promptly solving the crisis it inadvertently created by parachuting 14,000 live cats onto the island.

So what was the systemic failure? What was the systemic solution? Actually, nothing! Every solution was focused on the symptoms and not the problem. What resulted is what the systems dynamics community would call "fixes that fail." Let us look at another example of a disease that is the most commonly reported vectorborne illness in the United States, is the fastest growing infectious disease in the United States, and saw over 30,000 cases in 2011 according to the United States Center for Disease Control: Lyme disease. Lyme disease is caused by a bacterium, *Borrelia burgdorferi,* which was first identified in the United States in Lyme, Connecticut, in 1977. It can affect people of any age and is transmitted from the bite of an infected deer tick or black-legged tick, with the majority of the cases being in the northeastern region of the United States. While most cases are curable with antibiotics, if left untreated, it can cause meningitis, facial palsy, arthritis, or heart abnormalities. Typical symptoms are the development of a large circular rash around or near the site of the tick bite, chills, fever, headache, fatigue, stiff neck, swollen glands, and muscle and/ or joint pain. This can last as long as several weeks, and pain in the large joints may even recur many years later.

So what is the system that causes Lyme disease? What is the systemic nature of this system? What is the systemic solution? What most people understand about Lyme disease is that it is carried by ticks, who live in the forest, and we become infected after being bitten by a tick during a joyous joint through the woods and brush. But let us look at the bigger picture. Ticks are arachnids that live on the blood of mammals. These mammals can take many forms, but for most of their hematophagy activity, ticks chose rodents, for example, rats, mice, and squirrels. Thus, as the rodent population grows, so does the tick population, and consequently the number of cases of Lyme disease. So is the systemic problem with Lyme disease the rodent population? Not quite.

It has been well documented that Lyme disease can be linked to . . . acorns! Yes, acorns. Acorns are predominantly produced by oak trees every 2–3 years as a tough, leathery shell nut, which contains the seed of life of future mighty oaks. Oak trees are well-recognized symbols of strength and courage. For example, many seals of United States government agencies have oak branches under a bald eagle to demonstrate the agency and country's strength. Some oak trees can live to 650 years and can exceed 4 ft in diameter and 100 ft in height. So is this mighty and majestic perennial the systemic problem of Lyme disease?

Before we reach a conclusion, let us describe our system of interest further and define it simplistically, as shown in Figure 5.1. This figure clearly shows that our grand symbol of strength is the source of our discomfort. As the oak tree produces more acorns (assuming there has been adequate water, sun, and air), the rodent population grows because there is more bounty for their nourishment and thus there is an increase in rodent reproduction. With more rodents, the tick population grows because they too have a plentiful source of nourishment. With more ticks, we have a higher probability of being bit by an infected tick, and, thus, more cases of Lyme disease occur. So what is the solution?

If we go back to our story of the cats, we would think of solutions to this problem such as to use a tick repellent with DEET, to spray trees and shrubs with insecticides, to hunt more rodents, or our all-time favorite: to eat more acorns. The common, ordinary acorn is one of the ancient foods of mankind. The first mention of acorns for human consumption was by the Greeks over 2000 years ago, and some estimates show that humans have eaten more acorns than both wheat and rice combined at this time. By consuming the

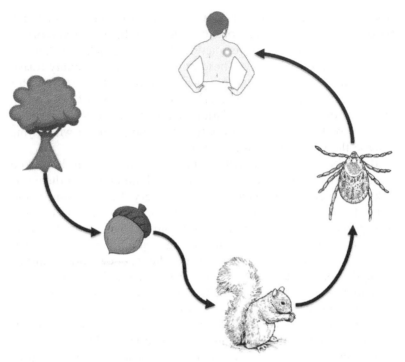

**Figure 5.1.** System of Lyme Disease

right amount of acorns, we could balance the system by keeping the reciprocating populations of the other parts of this system manage-able. Just see Grandpappy's Basic Acorn Recipes (http://www. grandpappy.info/racorns.htm) for some wonderful Pioneer Acorn Pancakes, which go great with a side of acorn grits and acorn biscuits and gravy.

All of these solutions seem probable, but they all result in the same "Operation Cat Drop" problem. For example, reducing the acorn population would reduce the rate of reproduction of oak trees, which give out oxygen; because of them we even have the rains, which nourish the Earth. With less rain, we would have less water for irrigation, resulting in less food production and leading to further famine. In short, oak trees maintain ecological balance, and reducing the number of oak trees may cause a further imbalance in the ecosystem. So what is the systemic solution? There is none. Eric Berlow (2010), renowned ecologist and educator says,

> if you want to predict the effect of one species on another, if you focus only on that link, and then you black box the rest, it's actually less predictable than if you step back, consider the entire system—all the species, all the links—and from that place, hone in on the sphere of influence that matters most.

But what matters most? What is the "sphere of influence"? What we learned about the two stories of infectious diseases is that the sphere of influence is not as simple as it first appears and that worlds of systems are more complex and tightly coupled. Yale University Professor Stanley Milgram's six degrees of separation is now more of a slang phrase than the results of a calculated experiment (Milgram 1967). Systemic thinking is as much about what happens between the parts as it is about what happens with the parts. Our natural tendency is to uncouple the system to understand it, but what we lose is the influence of the coupling. We have to learn how to understand the system without uncoupling it, moving from reductionism to wholism. We do not stand alone in our failures or solutions. In understanding the worlds of systems, we have to accept that how and why parts come together is as important as what the system does—relationships matter. Some have gracefully described this as what could be termed "cogeneration." Rachel Botsman (2010), in *What's Mine Is Yours: The Rise of Collaborative Consumption*, describes an emerging economy where trust and reputation are the real forms of currency; Martin Nowak and Roger Highfield (2011) explain in *SuperCooperators: Altruism, Evolution, and Why We Need Each Other to Succeed* that our ability to evolve successfully is significantly influenced by group selection and that cooperation is as much a part of evolution as mutation and selection; and Stephen Stearns, Yale University professor, explains in his lecture, "The Impact of Evolutionary Thought on the Social Sciences" (Stearns 2010) that

> there is a distinct possibility that humans are currently part way through an evolutionary transition between individuals to groups. The conflict between these two units of selection and levels of

organization, between biology and culture, may explain some of the tension in modern human life.

We have learned that diversity does not create stability. Injecting new elements into an existing system creates new connections that lead to oscillations in the parts and their relationships with other parts—this is why fixes can fail, or we can have unintended consequences. Diversity cannot assume that in belonging to the system, it will be accepted. Belonging involves a belief and acceptance that the parts are equal on some plane, yet diversity can create new and emergent behaviors, giving us systems we never had before—systems evolution.

In our search for the systemic nature of a system, we will find that our answers are in how and why the system comes together—its togetherness. Togetherness occurs when autonomous individuals commit to a collective vision, and the willingness to belong constitutes sacrificing individual objectives for a common mission. In togetherness, there is the envisioning of something superior that no individual or even collection of individuals could ever achieve. Separating out the individual is not feasible, and the associated pressing demands for its systematic solution rather than its systemic resolution is flawed. In togetherness, there is the idea of a structure that draws together these individuals, along with enabling resources, to serve a new higher purpose in addition to fulfilling their original terms of reference. To understand the systemic failure is to first understand the nature of togetherness.

We never contended that this book was about systemic solutions; it is about systemic thinking, and we seek to provide a way to look at systems that allows for understanding the problems so you can navigate the journey. Every great journey ends in a place you never knew it would take you when you started. But no journey ever started without an idea of where you are going and a map of how to get there. The next two journeys in this book will provide you with a way to frame the systemic idea and create your systemic map.

**JOURNEY II**

# SYSTEMIC IDEAS: THE CONCEPTAGON

# CHAPTER 6

# FRAMEWORKS

St. Mary's Church has stood in the English village of Kempsey, Worcestershire, for more than 800 years (see Figure 6.1 for a beautiful rendition of St. Mary's Church by Roger T. Cole). The chancel of the original aisleless cruciform building founded in the middle of the twelfth century was rebuilt a century later; at that time, the east window of five stepped lancets was also inserted. Later in the same century, the south aisle and arcade was added, the north aisle being added in the early fourteenth century. In 1759, a Musicians Gallery was erected at the west end, blocking the archway between the nave and tower, but this was removed in the Victorian era when modern pews replaced existing box pews. As successive generations of Christians followed their faith, the building in which they worshiped underwent gradual development. Here is an (adapted) extract almost as ornate in its eloquence as that part of the church's architecture it seeks to describe ("Parishes: Kempsey" 1913):

*Systemic Thinking: Building Maps for Worlds of Systems*, First Edition.
John Boardman and Brian Sauser.
© 2013 John Wiley & Sons, Inc., Published 2013 by John Wiley & Sons, Inc.

**Figure 6.1.** St. Mary's Church by Roger T. Cole

The tower is of three stages, with angle buttresses and an embattled parapet, having crocketed pinnacles at the four corners. The two-centered tower arch has flat paneled jambs and soffit, and the west window of the ground stage is of four large cinquefoiled lights with vertical tracery in the head. At the north-east is a blocked entrance to the vice, which is now entered by a modern external doorway. The bell-chamber is lighted on all four sides by windows of two trefoiled lights with traceried two-centered heads, and the ringing chamber beneath by windows of similar design on the north, west and south.

In the latter part of 2011, the tower of St. Mary's Church disappeared. It hadn't been stolen. It simply became obscured behind an ugly green makeshift curtain of hessian. This was draped around a skeleton of hollow steel tubes supporting a number of temporary wooden planked floors, which were decorated with ladders used rather precariously by stone masons as they undertook essential repairs to the building. In the mystery of the disappearing church, the prime suspect and sole culprit was scaffolding. Ironically, it was also that church's savior.

For many months, the familiar sight of the church's tower, normally quite visible over a few miles of the River Severn's valley plain, was replaced by a monstrous monument that stood out for all the wrong reasons. Yet this ugly ephemeral landmark existed for all the right reasons. Chiefly, it served to facilitate the essential reparation of a distinguished ancient monument for many good reasons; so that its legacy might be continued, its history vaunted through successive generations, and the ongoing worship by the faithful attendees securely located in a sanctuary protected from the elements and ravages of time. For a disagreeable period of time, the village played host to the good, the bad, and the ugly.

The church was good; its condition was bad and getting progressively worse year by year. The scaffolding was ugly; it seemed to make matters worse. And yet it was there to do good, to make the bad good so that the church could then continue its mission to make the bad good and the ugly beautiful.

And the point is? Scaffolding exists in many guises. The word has been adopted, some would say hijacked, by professions other than those who erect and maintain various buildings. Via its adoption, it has also changed its name. A term very familiar to legislators, politicians, economists, and engineers is *framework*. The idea behind this word is identical to the notion of scaffolding. Frameworks composed of principles, policies, rules and regulations, and indeed ideas themselves are used by a diverse group of professionals to enable real progress in their specific endeavors.

Latterly a breed of "new scientists," known as enterprise architects, has adopted frameworks with unreserved zeal to accomplish organizational reform and innovation. Engineers operating in the most advanced positions of product design have been using frameworks very successfully to ensure smooth progress in product development, manufacture, and testing. Using frameworks enables engineers to respond with extraordinary degrees of agility to the most dynamic environments so that products will emerge adaptively and stably regardless of changing requirements and uncertain operating scenarios.

The constitutional framework, which is the law as far as the United States is concerned, has demonstrated remarkable robustness and evolutionary capability for more than 300 years, making the United States a beacon of democracy to a world that hitherto has known only tyranny and feudalism.

Frameworks help build societies; they help deliver international agreements; they can produce new "buildings," which are the contractual arrangements between companies that provide financial stability to underpin risky product development, such as the Boeing 787 Dreamliner or the Trent 1000 engine that powers it. Frameworks are not the buildings themselves, but without them it's very difficult to make progress on raising new buildings and maintaining valuable ones for posterity. Frameworks absorb a massive amount of management attention, and one often wonders why. Is it time (effort and money) well spent? The answer has to be yes. So long as it is clear that the framework developers (and users) understand that their work is in support of the "buildings," the answer is most definitely yes.

We believe in the virtue of frameworks while recognizing that they are by no means the end product but are merely servants in the goal of real purposeful activity: the improvement in the quality of life for all mankind and, perhaps, the pursuit of truth. We know that quality of life, while enabled by the development of systems, is also inhibited by them. We also know that the pursuit of truth is a complex issue.

Our belief is that by understanding worlds of systems, those phenomena that emerge by virtue of the monotonically increasing population of systems, we can continue the process of improving the quality of life for all while reducing dramatically the negative effects that systems naturally introduce. We further assert that a better understanding of these worlds of systems will lead us more assuredly to a knowledge of the truth.

To back up our claims and lay bare our beliefs, we have chosen in this second journey of the book to introduce you to a framework of systemic ideas that we have found to be of practical value. This particular scaffolding of fundamental system concepts has helped us to "restore many churches." It consists of 21 basic ideas, all of which would be intimately familiar to professional systems thinkers and, for the most part, highly recognizable in common parlance. This makes them ideal for the spread of systemic thinking to a broad audience of concerned citizens. It is to the description of these ideas we now turn.

# CHAPTER 7

# THE CONCEPTAGON

Whether it is apocryphal or not, the story is told of how the famous movie mogul Samuel Goldwyn was once approached with an irresistible package deal to make a blockbuster movie with an all-star cast where none of the stars had ever previously appeared in a movie with any other of the stars. It would be a first for all of them. Winning trick or a risky business? Goldwyn knew his industry better than anyone. His judgment could be trusted. His conclusion was resolute. His response unequivocal. He told his importuners: "Include me out!"

This malapropism just described, which could easily form the basis of a pejorative attitude toward the legendary filmmaker, should not be so peremptorily scorned. Strangely, it conceals a hidden truth and one that systemic thinking can beneficially leverage. It speaks in coded terms of a number of fundamental system ideas, and we have embraced these in our scaffolding, the Conceptagon (Figure 7.1). The Conceptagon consist of seven triads,

*Systemic Thinking: Building Maps for Worlds of Systems*, First Edition.
John Boardman and Brian Sauser.
© 2013 John Wiley & Sons, Inc., Published 2013 by John Wiley & Sons, Inc.

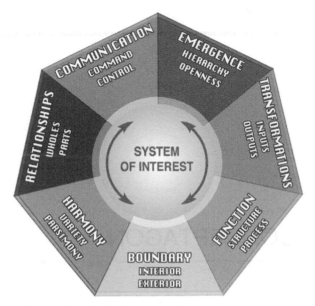

**Figure 7.1.** The Conceptagon

resulting from 21 concepts. The triad represents synthesis. Three lines are necessary to form a plain figure, and three dimensions of length, breadth, and height are necessary to form a solid. Thus, the triad is needed to form the whole of a systems concept. Seven represents wholeness or the completion of a cycle. Thus, seven triads make a whole or complete the cycle of the Conceptagon. Twenty-one or $3 \times 7$ represents the perfection by excellence or the harmony of creation. Thus, 21 becomes a natural product of the creation of the Conceptagon for us to better understand the harmony of systems.

So why 3, 7, 21? Why not? We become what is engrained in our DNA, so in the creation of the Conceptagon, it is self-evident our systems DNA would yield a framework that is a system in its own. Therefore, the concepts and triads do not stand alone. The linkages across triads form naturally, steering the user into new lines of inquiry as to what the system of interest really is in essence and how its detailed design and/or analysis must take into account new insights that the Conceptagon may reveal. We believe that the Conceptagon achieves two goals: first, it forms a basis for intelligent debate and effective collaboration between systems people of all walks of life, and, second, it provides a holistic view of the entire mission, ensuring that whatever specific pieces the specialists provide, the whole itself is coherent, efficient, and suited for purpose. This collection of ideas is both scale-free, covering multiple levels of systems effort, and horizontally integrative, uniting multidisciplinary labors at any given scale. Because there are no prescriptive methods for using the Conceptagon, users enjoy the freedom to think in new ways about their systems of interest, choosing their own navigation of the set of triads as it becomes obvious and intuitive so to do. In the remaining chapters of Journey II, we will travel through the Conceptagon.

# CHAPTER 8

# BOUNDARIES, INTERIORS, AND EXTERIORS

The most basic idea imaginable concerning any system is its *boundary*. Without appropriate demarcation, it's difficult to know what one is talking about, what it (the system of interest, or SoI) is opposed to and what it's not. And for that matter, where it is relative to its background and neighbors. Boundaries provide necessary distinction. They help you get hold of the system—both physically and cerebrally. Boundaries lend shape, surface (or superficial) appearance, and point to substance. By considering a variety of examples, we might acquire a firmer grip on the meaning and utility of this basic system idea.

The most obvious kind of boundary and simplest perhaps to appreciate is that of a physical boundary. Strong fences make good neighbors, as the saying goes. So private property is clearly delineated from that which belongs to another, and the legality of a property's physical boundary is often reinforced by rail and post fence or barbed wire if the property owner is resolutely determined

*Systemic Thinking: Building Maps for Worlds of Systems*, First Edition.
John Boardman and Brian Sauser.
© 2013 John Wiley & Sons, Inc., Published 2013 by John Wiley & Sons, Inc.

to keep out trespassers. So in addition to the basic purpose that a boundary serves, namely, that of demarcation, it can also be strongly associated with protection; in other words, boundaries might act as barriers that keep out unwanted entrants. They are in effect separators of the interior from the exterior, with the former being safeguarded from undesirable influences that exist on the outside while the latter is now regarded as space that cannot be similarly controlled because it lies beyond the powers, jurisdiction, and authority of the system basically beyond its reach. Of course, extending the system boundary increases the interior and gradually encloses what lies outside, which now comes under the system's control and adds to its resources, for example, in the case of a piece of property, shrubs, creeks, and pastures. We should observe therefore that a system's boundary need not be fixed but can flex, depending on a variety of factors.

The boundary not only might be flexible, it also might be fluid, by which we mean permeable. At first sight, this attribute might conflict with that of protection, but the fluidity or permeability would be under the control of the system whereby undesirable influences remain excluded whereas desirable inputs, such as additional resources, are permitted entry, and undesirable assets, such as waste material, nonconformists (in the case of social systems), and toxic assets (in the case of financial systems), are suitably disposed of. Intelligent permeability poses a management overhead for the system, but one that is usually worthwhile given the need for a system to exhibit openness, an attribute that allows it to closely monitor what is going on in its exterior and thereby maintain its strategies for survival in the first instance and prosperity thereafter.

What other purposes might a boundary serve in addition to protection and permeation? It can be productive by serving a purpose in its own right and on behalf of the system's interior and in association with its exterior. Take the iPhone as an example. Much is made of the industrial design element of the iPhone and of Apple, its creator. Yet the revolution that this platform wrought was achieved by making its systems boundary functional as both an input and an output device. The touch screen enables users to signal intentions to compose messages, search for contacts, download apps, and make phone calls. Gone are clunky keyboards with wasteful buttons. In are onscreen QWERTY displays with manifest confirmation of touch via feedback in sound, simulated clicks, and in sight, with ephemeral pop-up keys.

Perhaps the idea of using the boundary of a system as a crucial part of its functionality is borrowed by the digital world from the atomic world. After all, the walls of dwellings, an evident part of their system boundary, have for a long time been used as the home for conduit, running electrical services in stylishly unobtrusive fashion. So too have they been insulators of heat, keeping the home warm in winter and cool in summer. This feature may be argued to be a part of the boundary's protective or permeation purpose. Perhaps so. But when that insulation system has sensors attached that can control the amount of heat flow, in either direction depending on the relative temperatures between the outside and inside, it cannot then be denied that this boundary is adding to the efficiency of heating and cooling the home, making it a more cost-effective and therefore productive system.

These practical ideas for making the system better stem from a firm grasp of the concept of system boundary and the understanding that all systems are fundamentally constituted of an interior subsystem, an exterior subsystem (notwithstanding the reality that a system cannot control its exterior), and the boundary subsystem. This constitution is beautifully illustrated by the following puzzle involving three houses—A, B, and C—and three utilities—gas, water, and electricity. Each house needs a connection to each of the three utilities, but the puzzle insists that no connection between a house and a utility may cross another such connection. If we lay out the puzzle according to Figure 8.1, we observe that each home has two such connections without violating the constraint. Three additional connections are needed: A needs electricity, B needs gas, and C needs water. There are at first sight two "spaces" available. One is on the interior of the house–utility ring, the other on the outside. We have a system in place. Its interior and exterior are available for further service. But once they have been deployed, accommodating two of the missing three connections, that's it. There's no more space! If A gets its electricity from the interior space and B gets its gas from the exterior space, then

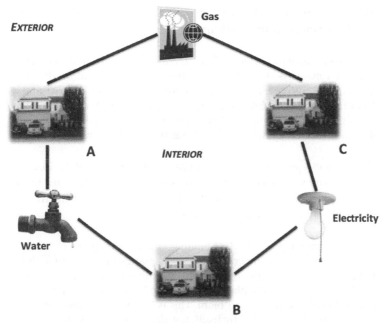

**Figure 8.1.** Puzzle: Boundary, Interior, and Exterior

C's connection to water cannot be made without crossing an exist-ing connection.

This puzzle is solved by observing that there is a third space. It is the system's boundary, which comprises connections, utilities, and houses. If C's water supply can come via a utility or a house, then it needn't cross a connection, and the puzzle is solved. This may not be possible in practice (both home and utility owners are picky about their individual boundaries!). But the puzzle did not impose this constraint. And the breakthrough comes from recognizing that a system boundary is a principle constituent of the system itself.

The beauty of systemic thinking is that these three subsystems are to the fore when considering any specific system of interest, and the payoff for giving them due consideration is the identification and management of systemic risk and the avoidance of solutions that compound the original problem.

What of nonphysical boundaries? Take a look at the text box below. Can you read the contents?

> The phaonmneel pweor of the hmuan mnid. Aoccdrnig to a rseearch taem at Cmabrigde Uinervtisy, it deosn't mttaer in waht oredr the ltteers in a wrod are, the olny iprmoatnt tihng is taht the frist and lsat ltteer be in the rghit pclae. The rset can be a taotl mses and you can sitll raed it wouthit a porbelm. Tihs is bcuseae the huamn mnid deos not raed ervey lteter by istlef, but the wrod as a wlohe.

We suspect you can. And yet you ought not to be able to, because apart from the words with three letters or fewer, all the other words are a jumbled mess. Somehow, it is possible for the brain to read this gibberish and make the following sense of it:

> The phenomenal power of the human mind. According to a research team at Cambridge University, it doesn't matter in what order the letters in a word are, the only important thing is that the first and last letter be in the right place. The rest can be a total mess and you can still read it without a problem. This is because the human mind does not read every letter by itself, but the word as a whole.

Amazing! As long as the boundary of the word is in place the interior, regardless of the jumbled mess that it is, can be correctly assembled. This is indeed a great tribute to our human minds. But it's also a feather in the cap of the system boundary!

What about a new twist for a boundary? Suppose a question were to be regarded as a boundary. Something along the lines of: "Who came after Harry Truman?" In what sense can this inquiry be viewed as a boundary? Well, first let's recall who Harry Truman was. For one thing he was the 33rd President of the United States of America. He succeeded Franklin Delano Roosevelt, who remarkably was elected for an unprecedented fourth term in 1944 as the World War II was drawing to a close. FDR died in office, and Truman, his Vice President at the time, became the new occupant of the White House and Commander in Chief as the Constitution requires. Truman himself was elected in 1948, but was beaten by Adlai Stevenson in the Democratic race for nomination in 1952. However, it was a Republican who became the 34th President.

This is all useful context, forming a kind of exterior to the system whose interior is then the answer to the question "Who came after Harry Truman?"

When we pose this inquiry to our students, many volunteer the answer Dwight Eisenhower, who indeed was Truman's successor as President. When we tell them that the answer we are looking for is Doris Day, most are bemused. Their temporary consternation is resolved by the magical powers of Google. When they enter "Harry Truman Doris Day" into the search bar, at the top of the list of Google's responses is: "We Didn't Start the Fire," a song written and recorded by Billy Joel whose opening lines are:

> Harry Truman, Doris Day, Red China, Johnny Ray
> South Pacific, Walter Winchell, Joe DiMaggio . . .

Billy Joel, who was a history nut and wanted to be a history teacher at one time, wrote the song in homage to the landmark events of his life (he was born in 1949). The song is a new context for our question and the answer is a different one to the question in which "comes after" means "succeeds as President of the United States."

This mildly entertaining distraction makes us revisit simple words like *boundary* and treat them with greater dignity. Were we not to, we might very well exhaust ourselves finding answers to questions that meant something other than what we took for granted. Wild goose chases, searches for solutions to the wrong problems, have been a salient characteristic of our increasingly complex world, one in which systemic failure appears inevitable. Taking a little time at the outset—what some call "putting pain in front of the activity," can make that activity hugely productive and avoid considerably worse subsequent pain than the little endured initially.

A final word on boundary as a concept we can utilize in systemic thinking. Charles Fine has defined a variety of proximities, alternatives to the familiar geographic one (Fine 1998). Among these are culture, organizational structure, and e-maturity. He develops these ideas in order to compare and contrast corporations that are conceivably "near" to one another without actually being neighbors in either the geographic sense or in terms of occupying the same industry landscape. Fine argues, for example, that corporations that have bureaucratic structures, such as the military or a federal agency, will be nearer to one another than to those with, say, a net-centric organization, in which the chain of command and reporting structures are less formal, lengthy, and time-consuming. Similarly, corporations that utilize near-identical IT structures will be closer, and therefore more capable of being immediately productive in any intended cooperative ventures, than the case where these IT services and platforms are very dissimilar. This notion of alternate proximities enables us to conceive of unusual boundaries defining systems in a novel fashion. Thus, comparison of systems proceeds on the basis of similar interiors, even though in other respects, the companies may very well perform different operations and occupy different industry spaces. Such comparisons form a useful preparation in establishing communications between systems, which may well be a precursor to mergers and acquisitions, an exercise that as we know has the effect of enlarging some and destroying other established boundaries.

What have we learned thus far? That as difficult as it might be to trace the boundary of any given system, apart from the obvious geographic ones, the idea is useful for many reasons. First, boundaries are strongly suggestive of preservation and protection—a shield for the system so that its survival might be safeguarded. Therefore, boundaries must be robust, capable of withstanding shocks and hazards that inevitably inflict the system. But while being sturdy, boundaries must also be flexible, capable of both contraction, without endangering system survival, and expansion, without overwhelming the system. Boundaries can also be productive in their own right, performing functions on behalf of the system and in harmony with the system's own ongoing activities. Boundaries must also be permeable, enabling the expedient ingress and egress of, for example, materiel and information.

Boundaries can exist in the mind of the observer as well as being "real." The imaginative construction or perception of a boundary enables us to appreciate, for example, the context of questions that are put to us. Understanding context is a great enabler of mediation and a powerful forestaller of premature action that so often proves nugatory.

The notion of boundary inevitably leads us to two other concepts, namely, those of interior and exterior. The former in effect represents system content—what the system possesses that it works with in order to perform its function or fulfill its purpose, that which must be protected by the boundary. Content is something over which the system can legally and practically exercise control. The exterior represents the environment of the system, about which the system can normally do very little, having neither direct authority nor control. Nevertheless, the exterior can, by various stratagems, be influenced. The degree of influence that can be exercised may be a determinant of contracting or expanding the system boundary.

These three concepts form a little family unit of their own. The exterior will include other systems, each having boundaries and interiors. So in crude terms, what is on the outside of a system is a bunch of other insides. And knowledge of these interiors can be shared only if boundaries are both protective and permeable. The more intelligent this combination can be made, the more stable will be this world of systems. Boundaries enable us to focus, therefore, on what primarily matters: the system's own content. Yet we need to be aware not only of that which lies within but also of that which is beyond us that might one day become part of us, or we it. Maybe Sam Goldwyn had it dead right without realizing the full import of his remark. "Include me out" means the boundary shows me what's important—everything! The fact that we can't execute this sentiment and must prioritize makes systemic failure more probable and systemic thinking imperative.

# CHAPTER 9

# PARTS, RELATIONSHIPS, AND WHOLES

When we venture through the curtain that is the system boundary in order to explore the interior, what do we find? To answer this question, let's take a look at three examples: the human heart, the Apple iPhone, and a book by the English author Charles Dickens, *Great Expectations*. As we journey, our goal is to make sense of everything we find at an abstract level, at what we might call the system level. It is then that we can become more confident of what *system* means and subsequently what its derivatives of systemic failure and systemic thinking mean. On this journey we will discover another abstract family of three units.

## ANYONE WHO HAD A HEART

The human heart weighs between 7 and 15 oz (200–425 g) and is a little larger than the size of a fist. By the end of a long life, a person's

*Systemic Thinking: Building Maps for Worlds of Systems*, First Edition.
John Boardman and Brian Sauser.
© 2013 John Wiley & Sons, Inc., Published 2013 by John Wiley & Sons, Inc.

heart may have beat (expanded and contracted) more than 3.5 billion times. Each day, the average heart beats 100,000 times, pumping about 2000 gal (7571 L) of blood.

Located between the lungs in the middle of the chest, behind and slightly to the left of the breastbone (sternum), the heart is surrounded by a double-layered membrane called the pericardium. The outer layer of the pericardium surrounds the roots of the heart's major blood vessels and is attached by ligaments to the spinal column, diaphragm, and other parts of the body. The inner layer of the pericardium is attached to the heart muscle. A coating of fluid separates the two layers of membrane, letting the heart move as it beats, yet still be attached to the body.

The heart has four chambers. The upper chambers are called the left and right atria, and the lower chambers are called the left and right ventricles. A wall of muscle called the septum separates the left and right atria and the left and right ventricles. The left ventricle is the largest and strongest chamber. The left ventricle's chamber walls are only about half an inch thick, but they have enough force to push blood through the aortic valve and into the body.

Four types of valves regulate blood flow through the heart: the tricuspid valve regulates blood flow between the right atrium and right ventricle; the pulmonary valve controls blood flow from the right ventricle into the pulmonary arteries, which carry blood to the lungs to pick up oxygen; the mitral valve lets oxygen-rich blood from the lungs pass from the left atrium into the left ventricle; and the aortic valve opens the way for the oxygen-rich blood to pass from the left ventricle into the aorta, the body's largest artery, where it is delivered to the rest of the body.

Electrical impulses from the myocardium cause the heart to contract. This electrical signal begins in the sinoatrial (SA) node, located at the top of the right atrium. The SA node is sometimes called the heart's "natural pacemaker." An electrical impulse from this natural pacemaker travels through the muscle fibers of the atria and ventricles, causing them to contract. Although the SA node sends electrical impulses at a certain rate, your heart rate may still change depending on physical demands, stress, or hormonal factors.

The heart and circulatory system make up the cardiovascular system. Working as a pump that pushes blood to the organs, tissues, and cells of your body, blood delivers oxygen, and nutrients are delivered to every cell while also removing the carbon dioxide and waste products made by those cells. Blood is carried from the heart to the rest of the body through a complex network of arteries, arterioles, and capillaries and is returned to the heart through venules and veins. This vast network has an end-to-end length that would circle the earth twice, with some to spare!

A heartbeat is a two-part pumping action that takes about a second. As blood collects in the upper chambers (the right and left atria), the heart's natural pacemaker (the SA node) sends out an electrical signal that causes the atria to contract. This contraction pushes blood through the tricuspid and mitral valves into the resting lower chambers (the right and left ventricles). This part of the two-part pumping phase (the longer of the two) is called diastole. The second part of the pumping phase begins when the ventricles are full of blood. The electrical signals from the SA node travel along a pathway of cells to the ventricles, causing them to contract. This is called systole. As the tricuspid and mitral valves shut tight to prevent a backflow of blood, the pulmonary and aortic valves are pushed open. While blood is pushed from the right ventricle into the lungs to pick up oxygen, oxygen-rich blood flows from the left ventricle to the heart and other parts of the body. After blood moves into the pulmonary artery and the aorta, the ventricles relax, and the pulmonary and aortic valves close. The lower pressure in the ventricles causes the tricuspid and mitral valves to open, and the cycle begins again. This series of contractions is repeated over and over again, increasing during times of exertion and decreasing while at rest, when beats are normally about 60–80 times a minute.

While marveling at the interior operations of the human, we should observe that this ingenious system does not work alone; it has an exterior face. We have already seen how it forms part of a wider cardiovascular systems. Not only so, the brain tracks the body's environmental conditions, including climate, stress, and level of physical activity—and adjusts the cardiovascular system to meet those needs (Texas Heart Insitute 2012).

While marveling at this organ, how can we make sense of it at the system level? Our journey was essentially one of decomposition. In other words, we wanted to know what comprised the heart—what its parts were, and how these parts were related both spatially—as a structure—and dynamically—the parts being linked by some repetitive, cyclical process. What parts and relationships do we find? The pericardium is a kind of system boundary, protecting the heart while enabling it to act as a pump and yet remain attached in a fixed location. In fact, this system boundary has three parts, with the myocardium being the inner layer of cardiac muscle, which exercises the pumping action under electrical impulse, and the endocardium forming an inner lining. Space is an important part of the heart. To be technically precise  four chambers: two atria, collecting spaces, and two ventricles, resting spaces. Valves, four of them, are crucial parts governing the timely flow of blood between chambers. With the aortic and pulmonary valves closed, the tricuspid and mitral valves open under the contraction of the collecting chambers forcing blood into the lower resting chambers. With the ventricles full of blood, the tricuspid and mitral valves shut tight to prevent the backflow of blood, and the aortic and pulmonary valves open wide to support the flow of blood to the lungs for oxygenation and to the rest of the body for the oxygen-rich blood via the aorta. A syncopation to admire with parts in a uniquely harmonious relationship that are in principle sustainable for more than a century. Can Apple beat that?

## HANGING ON THE TELEPHONE

Steve Jobs had magical powers. On the day he rejoined the company that he had helped found two decades earlier (on April Fools' Day 1976) and from which he had been rudely and errantly ejected in 1985, Apple had 90 days' reserve of cash. On the day he died (October 5, 2011), the rejuvenated company had more cash at its disposal than the U.S. federal government.

Jobs' magic was not reserved for corporate resurrection. It was something for the world to behold, most impressively when he imperiously commanded the stage, garbed in his trademark black turtleneck, blue denim jeans, and New Balance sneakers, thrilling a global audience with news of yet another "amazing" product.

After interminable rumors and with expectations running at fever pitch, Jobs announced on January 9, 2007, how the iPod, a breakthrough Internet device (e-mail, browsing, and other services) *and* a revolutionary mobile phone had finally come together. In one box. The iPod, a breakthrough Internet device, and a revolutionary phone. Not three devices. Just one box. The iPhone. Breathtaking is not a big enough word to express that news and what followed. What followed of course was phenomenal sales growth, the demise of the dumb cell phone, and an entirely new world that none but this extraordinary visionary had foreseen and from which none would ever choose to exit.

In the 5 years since that breakthrough moment, several versions of the iPhone have appeared. What has been constant throughout is Apple's unswerving commitment to smart technology, super-friendly usability, impeccable industrial design, revolutionary user interfacing, proprietary hardware design, and the best operating system design that portable computing has ever known. The constant has been clear architecting principles and the most powerful architectures possible. With this under fixed control, it's no surprise that peering through the curtain of the iPhone to explore its interior yields a list of coherent parts, many of which are sourced reliably and cost-effectively by multiple suppliers, all of them seamlessly integrated via robust relationships predetermined by Apple, yielding an emergent whole experience for users that sets the bar for all others. So by surveying what the tech-savvy crowd had to say about the iPhone 4S, relative to its predecessor, let's take a closer look.

Our generic guide is to figure that whatever detail lies within, we expect to come across these key items: a processor (formerly central processing unit, or CPU), a graphics processor (or GPU), memory, camera, antenna, chassis, the odd gyroscope and acceler-ometer, a display, and maybe a few unusual goodies. We can't see the software, but it's there: OS X parachuted in from the Mac.

So what Apple did, primarily, is to make over the inside of the iPhone 4. The 4S has an A5 processor, as expected, bringing the same dual-core processing power that the iPad 2 sports to the iPhone. Alongside that comes a better GPU, which Apple says provides "up to seven times faster graphics."

By combining the experience from the iPhone 4 and the CDMA edition, Apple has made the 4S a world phone, with both CDMA and GSM capabilities, requiring some clever antenna action as well as some tweaked internals. As well as ensuring good performance, from a year of field experience, this means Apple controls production costs downward, since they're making only one device. Apple say the radio systems of the new phone can cope with 14 Mbps download data rates, meaning that it's almost approaching some 4G speeds even though it's definitely a 3G-tech device.

The camera inside the 4S was also given some attention by Apple, and the choice of hardware showed a real understanding of what the photography game is all about—it's not the megapixels. Pixel count is just a part of the game, and the optics and other aspects of the camera, such as Apple's choice of a back-illuminating sensor with a design that means 75% boost in light performance, matter most. The tiny lens on the unit now has five optical elements, welded together to optimize performance so that Apple says it has "30% more sharpness" over the iPhone 4, with an improved IR filter for better color reproduction. The physical changes mean the iPhone 4S's camera can handle low-light situations much better than its predecessor (including incorporating a dynamic noise suppression system).

On a more mundane level, Apple kept the iPhone 4's chassis design the same. There's a good reason: it works. Apple's been making these things (or at least its Far Eastern suppliers have) by the tens of millions for well over a year now, and the process has been optimized to the point that the cost of building them is tailing off, fast boosting profits. The decision also makes for improved reliability, allowing Apple to avoid an unenviable return to "antennagate," learning the lessons about making the radio system of the iPhone 4S "just work."

Apple tweaked the hardware–software interface for the device, meaning image captures take 33% less time, and takes just 1.1 seconds after activating the camera app to allow you to take a photo, beating the pants off peers like the Samsung Galaxy S2 and the Droid Bionic. The iPhone 4S also has improved image data transport and processing time, so it takes just 0.5 seconds to take the next image, compared with the S2's 1.3 seconds, and the Bionic's 1.6 seconds. Add in 1080-pixel video capture, image stabilization in video mode, and face recognition, and the camera really threatens traditional point-and-shoot cameras. The process to tweet out new images is just a single tap at the screen, and with iCloud, the photos are automatically shared to a backup, features most cameras can't match.

Then there's Siri, a transformational voice-controlled personal assistant you can call up anytime by holding down the home button for a few seconds. You can ask it the weather or time, get directions from Yelp, schedule a meeting on your calendar, reply to messages, play a song from iTunes, or ask any factual question via Wolfram Alpha—all with voice commands. Voice processing happens in the phone, unlike Google's equivalent services, which send samples of sound off to the cloud for processing, which takes time, requires a good 3G connection, and hits users on limited data tariffs. Siri may possibly become one of the biggest draws to the iPhone 4S.

In summary, the iPhone 4S is very similar in external design to the iPhone4, but on the inside, Apple has honed, polished, boosted, adjusted, extended, and pushed the capabilities of the iPhone 4S to the max. And that progress shows us a trajectory for future events. Or maybe not?

We've looked at a lot of detail, and that's certainly one way to break out a set of parts and their interrelationships. But we could go deeper, for example, into the intricacies of the A5, appropriately termed a System on Chip (SoC), or the inner workings of the OS X, the outstanding operating system that powers the Mac products and latterly the iPad and iPhone.

Alternatively, we could step back and adopt an entirely orthogonal perspective. We could choose to regard the ecosystem of apps developed for the iPhone as a crucial part (over 5 billion downloads to date), together with the range of services or features that support users' real-world needs. Then we could aim to discover how apps, services, the hardware platform, and iOS interrelate to provide user experience.

Envisioning parts and relationships (and wholes) calls for an architectural view and is one that can be decided by the observer with a specific perspective in mind. Whatever the view, it only adds to our amazement in the case of the Apple iPhone. It may not be the human heart, but as an essential accompaniment to social existence and as an example of "intelligent design," it's as good as it gets. As for tomorrow, who knows?

## WHAT THE DICKENS!

A young boy helps an escaped convict whom he meets on the moors by bringing him some food. The convict is recaptured and hauled off to Australia. The boy is an aide to a blacksmith. His prospects are defined by completing his apprenticeship and some-day taking over the business from his mentor, the widower of his elder sister. The boy has no expectations.

He is given an opportunity to serve on a part-time basis at the mansion of a wealthy lady, a jilted bride wracked by pain, haunted by failure, and energized in her aging years by revenge. The boy serves well and is praised. His reward is to behold the youthful beauty of the lady's ward whom she is infusing with her cold-hearted spirit of revenge. The boy has few expectations.

A solicitor brings the boy news from out of the blue. The boy is to receive money from an unknown benefactor, enough for him to be schooled in London and prepared for a gentleman's life. The boy now has great expectations. And what happens to the boy, the girl, the lady, the smithy, and the benefactor is told in incomparable style by someone universally acknowledged to be one of the world's greatest authors.

The story is described in the book *Great Expectations*, by Charles Dickens. Can this book be considered a system in its own right? If so, what are its parts and relationships? What exactly is this whole? And how can our findings be compared with those of the human heart and the Apple iPhone? These eclectic findings will help us better understand the meaning of *system* and thereby systemic thinking.

The book *Great Expectations* is in one sense just like any other book. Its parts consist of pages of paper containing patterns of ink and bound together with a cover. Paper, ink, cover, and the glue that adheres the pages are the parts. The relationships are the patterns of ink on the printed pages, the order of the pages that enables narrative flow, and the binding that holds the cover and pages together. That's it—or is it? The constituency with most interest in this system are the printers, the publishers, and the readers who don't want the book to fall apart while it's being handled, nor the story to fall apart because a page was printed back to front. There are other constituencies with different perspectives who will therefore have a different view of the system.

One such constituency is the literature class that studies the book for the purposes of deepening its members' understanding of the story and of storytelling. To this group, the parts may be the characters, including Pip (the boy), Joe Gargery (the smithy), Abel Magwitch (the convict), Miss Havisham (the jilted bride), Estella (the ward), and Mr. Jaggers (the solicitor). The relationships are formed by the characters intertwining via an elaborate drama climaxing in the revelation of who is really who (the mysterious benefactor and the father of the ward, for example). The whole is the experience that readers have as a consequence of encountering the characters and their various relationships. This is the same system seen at a different level. Are there different constituencies and yet more levels?

Consider a group of actors who are being considered to play the characters in a production of the book. What is their interest? For sure they will need to know the book and far better than a literature major. The reason for this is that when they play their parts in the production, they will want to bring the characters to life, making them credible not only in ways that are true to Dickens' intentions but also in a way that responds to the culture of their own age. This is a new interest group with a different perspective and a fresh interpretation, making sense of the same system and in so doing creating richer parts, deeper interrelationships, and a new whole.

Last, philosophers and critics, and perhaps some politicians, will analyze *Great Expectations* from the perspective of human nature, of social norms and human values, of ethics, morality, and psychology. For them, the characters and their relationships are merely props with which to lay bare the traits of human kind: aspiration and fear, love and rejection, justice and retribution, envy and greed, mystery and surprise, hope and reality. For the actors, the props they needed to bring the characters to life were the lines of dialog and the events that unfold. For the readership, the props were yet more elementary: ink, paper, and glue.

Different groups, each one searching for their "building" and finding their individual whole, yet all needing these props, their particular scaffolding. As we engage in this kind of discourse, it is then that we find ourselves at the portals of systemic thinking.

## DON'T GET RID OF ALL THE PHONE BOOTHS!

Clark Kent is a mild-mannered reporter. Superman is the man of steel. The local phone booth is the link between the two. With the surge in demand for the iPhone (and its peers), this relic of a bygone age is threatened with extinction. It's an essential device for transforming a bumbling, shy innocent who knows what is good, into a gravity-defying, indefatigable superhero who does what is good. Surely, its eradication cannot be risked? By all means, let us enjoy our inalienable right to pursue happiness. Let us have our magic wands that bring the world to our pockets and enable us to tell the world what we think. But not at an impossible price: the end of the American way of life.

The two happy families we met in Chapter 8 and in this chapter emphasized a structural way of thinking. In other words, we wanted to know "What does this systems consist of?" What is inside it, about which we can make clinical decisions, and what is left outside, about which we can do very little—although we can exercise some influence? We wanted to know how the things that make up the system were connected. Our line of thinking, in other words, was compositional (or decompositional).

What we do know is that whatever the structure of the system turns out to be, it works. It hangs together. It's what gives the system its overall shape, substance, and stability. It makes the system the whole that it is. And the system is continually in control of this structure, which is just as well, for if it were not, it would fall apart.

But the notion of system boundary introduced us to the thought that there are things beyond its control, things that lie outside the system, and our concern should be how the system can continue to enjoy stability in light of events, circumstances, and influences that can cause injury to the system, possibly of a fatal nature. The structural nature of the system exterior is largely inaccessible to us, and the relationship that the system has with its exterior is at best a mostly unpredictable series of mysterious events and might at worst be unilateral. The way we express this technically is to say that the system's relationship with its exterior, and it with the system, is dynamic and that the perpetual search for a definable structure is subject to a dynamism that we need to understand.

So, in the next two chapters, we meet two new families that try to capture this thinking about dynamism. Unsurprisingly, they introduce ideas to do with *flow*, for example, of materiel and information (atoms and bits); *feedback*, in which circular flows can be leveraged to understand dynamic stability and equilibrium (or lack of it); and *fostering*, by which we can make changes to system structure and to whatever we believe to be the system's relationship with its exterior that we consider will be for the good of both the system and its exterior.

These next two chapters are therefore all about change and transformation, about how systems are constantly exchanging information and materiel with other systems and the environment, and in the process effecting change on what flows into the system and produces subsequent outflows. It's about how systems monitor these flows and regulate them so that they are not overwhelmed or become unstable. And it's about how systems, being faced with a constantly changing environment, might themselves need to undergo structural change in order to survive.

Systems produce change and undergo change. In that sense, they are transformational, they are like the local phone booth. We need them. Because, for sure, there are loads of Lex Luthors lurking.

# CHAPTER 10

# INPUTS, OUTPUTS, AND TRANSFORMATIONS

Our next family is made up of three ideas that are logically connected, and they are *inputs, outputs*, and *transformations*. Let's use these familiar ideas with respect to our very good friend, the human heart.

Blood flows into and out of the heart. It flows in from the rest of the body carrying a supply of carbon dioxide, which has been made as a consequence of the body using oxygen to supply vital resources to the muscles, especially our brain. This blood is then pumped out from the heart to the lungs as part of a process to remove that carbon dioxide, via exhalation, and to replenish the blood with oxygen, via inhalation. The heart must pump and the blood must flow, or this reoxygenation process will fail and the body will expire. The oxygenated blood flows back to the heart and is then pumped from there around the body in order to begin a fresh cycle of energy supply to our bodies. As far as this system is concerned, inputs, outputs, and transformations are pretty vital.

*Systemic Thinking: Building Maps for Worlds of Systems*, First Edition.
John Boardman and Brian Sauser.
© 2013 John Wiley & Sons, Inc., Published 2013 by John Wiley & Sons, Inc.

The cardiovascular system is at this point in time our system of interest. That is, if you like, it is our St. Mary's Church. But if we are to understand it, which will ultimately mean to manage it, *as a system*, we need some scaffolding, some system ideas, with which we can better appreciate its structure and its dynamics.

One thing we cannot fail to notice, then, using our new family is that some inputs are outputs from elsewhere and some outputs are inputs that go elsewhere. And this coidentity of an output being an input is the basis of a syncopation that is vital to the world of systems that make up our entire bodies and perhaps even make us who we are. So, for example, the oxygen-rich blood collected in the left atrium is an input from the pulmonary veins, coming from the lungs. It becomes an output to the left ventricle and an input, as far as this chamber is concerned, from the left atrium. This same blood subsequently becomes an output via the aorta and simultaneously a vital input to the rest of the body.

On "the other side" of the heart, blood that is rich in carbon dioxide is input to the right atrium via the superior and inferior vena cava from the rest of the body. Having been collected in this chamber, it become its output and identically the input to the right ventricle when it is later pumped as output to the lungs via the pulmonary artery where it is identically an input to the lung's oxygenation system.

Man did not design this system. Some believe in intelligent design and a divine creator. Others argue that it evolved and self-organized. But whichever camp he subscribes to, man is at least smart enough to emulate these natural (or supernatural) exemplars whenever he does design systems. However, it is not always possible to think of everything, and one example is worthy of note.

During the Apollo 13 mission, the explosion that caused the premature demise of the Odyssey meant that the Lunar Module (LM) had to become a lifeboat, and thanks to some brilliant engineering by the Grumman Corporation this proved possible. However, there was a snag. The removal of carbon dioxide for both the LM and the Command Module (CM) was dealt with by canisters of lithium hydroxide. For the LM, these were round cartridges, but for the CM, built by a different company, they were square. The supply of cartridges for the LM was inadequate for the lifeboat mission. Yet those from the CM were unusable because of their "incompatible" shape.

The lifeboat rescue was a piece of brilliant improvisation that leveraged outstanding systems engineering. But no one had anticipated this rescue at the level of detail that extended to asymmetric inputs and outputs. Thankfully, the folks at Houston came up with a fix using only materials known to be available to the Apollo 13 crew. Engineers on the ground improvised a way to join the cube-shaped CM canisters to the LM's cylindrical canister-sockets by drawing air through them with a suit return hose. This jury-rigged device became famously known as the mail box, a transformation of value at least equal to Clark Kent's local phone booth!

While outputs are identically inputs, inputs themselves undergo transformation into outputs under some regulated process or dynamic. All such transformations add value, but there are a variety of types. The blood in the left atrium undergoes a mild transformation. First, it is at rest in the chamber. Then, coming under pressure as the chamber contracts from the influence of electrical stimuli originating in the sinoatrial node, it emerges at speed. Same blood, new velocity.

In the lungs, the transformation is a little more dramatic with blood undergoing a change of content, unwanted carbon dioxide being replaced by life-giving oxygen. Likewise, in the muscles of the body, an inverse change occurs with precious oxygen and nutrients being consumed to power the body's function, and carbon dioxide and waste material being generated and collected in the blood that returns via the vasculature highways to the heart.

The nutrients spoken of earlier that are carried by the oxygenated blood enter the body by the digestive system. Inputs here are what we commonly refer to as food, for example, meatloaf. Thinking of this might make you hungry, but for now, we want to give you food for thought!

The creation of a meatloaf is an elaborate process. At a high level, it involves ingredients, a recipe, a stove, and some kitchen equipment. That's as detailed as it gets for us! What we do know is that the ingredients undergo a dramatic transformation when you follow the recipe, and the output, the meatloaf suitably glazed, bears no resemblance to the raw materials that were the original inputs. It's almost as if the inputs have been destroyed and an entirely new thing has emerged.

What is more, this output disappears—more correctly, is transformed—before it becomes the input we call food. First, it is cut into slices, and trying to piece them back together will never make the original meatloaf. Then the slices are served on a plate alongside maybe some mashed potato and steamed broccoli. And now for dinner! What happens once we've chewed on this awhile is another story. You're in a better position now to tell that story, with this third family serving as a prop.

This family provides a helpful introduction to the notion of flow that is certainly part of the dynamism that characterizes systems. But we can also let it provide us with insights into a second aspect of this dynamism that is termed feedback, an idea that has both positive and negative connotations, as we shall now see.

Let's consider a fictitious company called Tangerine, which decides to compete with Apple in the smartphone market. Initially, the demand for their rival product, the tPhone, is small because that's how things start out. But eventually, word catches on that this device is comparable to the Apple iPhone, in terms of functionality, but it beats the pants off it on price. Demand for the tPhone increases. As this demand increases the sales revenue into Tangerine increases. As the company's coffers swell, more money is available to expand the market, thereby making many more people aware of tPhone's virtues. Accordingly, as the market is expanded, the demand for tPhones increases, and round and round we go. This is a virtuous circle. It expands positively without limit and rapid growth is depicted by this endless feedback cycle.

Life isn't like that, right? What are the limits to growth? Well, as good as the tPhone is, the lower cost to customers comes at a price. Things go wrong and customers need help. Tangerine understands this and they have a customer service department to deal with concerns and to apply fixes. The problem is that as the market expands, the demand on the customer service increases, and the ability of that unit to provide a high-quality service degrades. Word soon gets out that the fixes take a long time, too long, and people are put off buying the product. Demand falls off in a negative

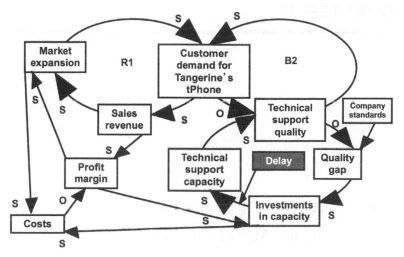

**Figure 10.1.** Tangerine tPhone Simultaneous Goals Causal Loop

feedback cycle. This does not have to be the end of the road for Tangerine, though.

They can set some quality standards and monitor the quality of service provided by the customer care department. By comparing what they measure with the standards required, they can decide which additional resources to apply to the care function so that it can improve the quality of its service and thereby restore the market's confidence that, although occasionally tPhones will need attention, all will be well. A causal loop diagram that depicts flow and feedback showing how the salient variables interact is given in Figure 10.1.

# CHAPTER 11

# CONTROL, COMMAND, AND COMMUNICATION

Mention of the feedback idea provides a useful link into our second family that helps us understand a system's dynamics, and this unit comprises the three concepts of *control*, *command*, and *communication*.

## CONTROL

Feedback is the basis of system regulation or what is better known as system control. This is the name given to the means by which a system governs and maintains its structure in order to keep itself stable and/or to allow it to pursue its goals. Control is a very simple idea that involves monitoring system status and taking appropriate action based on the observed status. Take, for example, driving an automobile down a highway at a constant speed. Suddenly, you are confronted with a sign indicating a maximum speed limit of 30 mph.

*Systemic Thinking: Building Maps for Worlds of Systems*, First Edition.
John Boardman and Brian Sauser.
© 2013 John Wiley & Sons, Inc., Published 2013 by John Wiley & Sons, Inc.

You monitor your speedometer and observe that your speed is 40 mph. Too fast! You take your foot off the gas and the vehicle slows down. It reaches 25 mph. Too slow! You mash the accelerator—gently—and observe that your speed becomes 30 mph. You've hit the target and take no further action. Either you do this or your cruise control does it for you. This is a case of negative feedback. It is the kind of action we saw Tangerine take relative to maintaining quality standards in its customer care department.

There's another kind. Apples that ripen on a tree give off ethanol, which is a substance that induces further ripening. As the ethanol makes its way through the air, it reaches other apples and stimulates their ripening. This in turn produces more ethanol, which accelerates ripening of all the fruit on the tree. Bumper harvest, on time. This is positive feedback. Now positive feedback would be inappropriate in motoring through a busy city since it would produce gridlock via stopped vehicles, or casualties via speeding cars. But it is appropriate to Tangerine growing its business rapidly by plowing back increasing sales revenue via increased demand for tPhones into additional marketing effort. This is the virtuous circle example of positive feedback, whereas it would be a vicious circle when applied to motoring in busy cities.

Control is a rudimentary component in technology. System Control and Data Acquisition (SCADA), for example, is part of the very fabric of all industrial control systems that monitor and control industrial, infrastructure, or facility-based processes. These include manufacturing, production, power generation, water treatment and distribution, wastewater collection and treatment, oil and gas pipelines, electrical power transmission and distribution, wind farms, civil defense siren systems, large communication systems, HVAC, and energy consumption.

## COMMAND

The field of systems dynamics has sought to extend the reach of control concepts from their natural home in these technology-based systems to other fields that are more societal-based, including the regulation of financial markets, the governance of the

Afghanistan conflict, and population control stratagems in developing countries.

But given the essential nature of human beings to act in irrational and perverse ways, it is unsurprising that, as much as this field has cast light on many social issues, it has not successfully ported all of the sophisticated algorithms that work a treat in technology to the control of human behavior. It seems that the bunch of atoms that we are achieves an emergent behavior in which conformance is not as evident an attribute as we find it to be in individual atoms.

In *Great Expectations*, Miss Havisham (the rich lady) tried to control Estella (her ward) and was successful only up to a point. This eventually rebounded on her. Later she tried to control Pip (the young boy), but once again with less than complete success. Control can legally and reasonably be exercised over individuals and groups of people when the controlled have given their a priori consent voluntarily. This applies in the military and in many employment situations, where senior officers and bosses can instruct subordinates to follow certain rules and methods with confidence. This we call command. But there are cases when even this regime fails.

In the movie *A Few Good Men*, Aaron Sorkin weaves a brilliant tale of how two enlisted marines carry out a command handed down to them from the company commander. These instructions, a code red variety, were to rough up another marine who was finding life in general and training in particular arduous and unappealing as a means to get the "weakling" to shape up. Regrettably, this person died as a result of what the two marines carrying out the code red did to him, and they were subsequently tried before a court martial on charges of murder, conspiracy to commit murder, and conduct unbecoming of a marine. The movie became a contest between the defense lawyer (played by Tom Cruise) and the company commander (played by Jack Nicholson): two Hollywood heavyweights slugging it out for the truth, however hard that is to handle. The final verdict exemplifies which commands are valid and which can overrule orders when they are morally suspect. In fact, Sorkin neatly covers himself by ensuring to mention in the movie that the company commander himself was ordered to desist from code reds, thereby restoring the moral authority of military command.

Admiral Horatio Lord Nelson was a military man and understood the importance of command and the need for orders to be obeyed. In this regard, he was a brilliant strategist. Moreover, he was an outstanding innovator and did much to enhance the virtues and extensibility of command by leveraging three key principles. Courtesy of the excellent commentary of Leigh Kimmel,[1] we are able to describe these principles at work in the life of one of England's heroic figures via the following extracts:

> The idea that an individual commander as the man on the spot should have the flexibility to deal with the situations as they came was a central part of Nelson's battle doctrine. He had a talent for communicating his ideas and plans to his captains so well that they understood what he would want them to do in any specific battle situation and carried it out as well as though he were there. Thus he was able to keep his orders general. In his orders for the assault in Tenerife, item six noted that his captains were "at liberty" to send more men and to land under Troubridge's direction rather than have to get specific orders from Nelson. He also had the battle plan for the Nile worked out almost two months before he actually entered Aboukir Bay. His was the master plan and he left the details to individual captains, believing that they had the good common sense to innovate and act independently. Foley's decision to go inside the French line at the Nile when he saw the opportunity fits perfectly with Nelson's philosophy of independence of command. Nelson's orders to his captains at Copenhagen were also quite bare and simple. He expected them to apply these general details to the specific situations they encountered as the battle unfolded. Finally, Nelson's famous memorandum circulated before Trafalgar gave Collingwood full latitude to fight his whole line as necessary.

> The concept of **creative disobedience** flowed naturally from his philosophy of independence of command. If his subordinates should have the freedom to deal with situations as they came up, he should be able to take the initiative as a subordinate in a battle, even if it meant ignoring orders. The first great example of this was his "famous indiscipline" at the battle of Cape St. Vincent, where he pulled out

[1]See http://www.leighkimmel.com/writing/academicpapers/nelsonsea.shtml.

of the line of battle in order to interdict the Spanish flagship, thus allowing the rest of the British fleet to catch up and get into fighting position. However this involved breaking the standing orders that no ship was to leave the line of battle without permission from the senior admiral. Oliver Warner claims that no other subordinate officer has taken such an initiative as Nelson did at Cape St. Vincent, although he does not make clear whether he is comparing Nelson only to other officers of the Royal Navy or officers of all navies (which would also require examining the history of the Pacific Fleet in World War II, which had its fair share of gung-ho admirals who didn't always mind Nimitz). Sir Nicholas Harris Nicolas, editor of Nelson's letters, suggested that Jervis didn't praise Nelson for his success because Calder, Jervis' flag captain, pointed out that Nelson had disobeyed standing orders to stay in the line of battle and that praising such disobedience would set a bad example for future officers. However Jervis is recorded as having responded to Calder's criticism with the remark, ". . . if ever you commit such a breach . . . I will forgive you also."

**Reciprocal loyalty** is the idea that one must give loyalty down the command hierarchy in order to gain true loyalty (as opposed to obedience through fear). Nelson seems to have understood this instinctively, although his year of service on a merchant ship at the very beginning of his career may well have helped to shape that instinctive understanding into practical action. His career shows many examples of the way in which he stood by his subordinates and saw to their welfare. Near the end of his years ashore, between the time of the storming of the Bastille and the execution of Louis XVI, certain elements of English society were becoming restive with the possibility of freedom promised by the French Revolution. While many aristocrats and country gentry were responding with hysteria, Nelson started looking for the cause of the problem and its solution. He went around the Norfolk countryside talking to ordinary people about their grievances and put the knowledge he gained to work. He also did his best to improve conditions aboard his ships and to see to the welfare of the sailors under him and their families. After the Battle of the Nile, Nelson wrote a letter to Lord Spencer (then First Lord of the Admiralty) asking after the welfare of the fourteen-year-old eldest son of a Marine officer who was killed aboard his flagship in that engagement. After the Battle of Copenhagen he wrote several letters trying to get recognition for his brave followers. In a letter to St. Vincent he expressed his belief that the commanders at Copenhagen should be given medals. In a letter to the Lord Mayor of London he claimed that he wouldn't complain if his reputation were the only thing involved, but he had the bravery of his subordinates to consider and wanted them recognized. And shortly before the Battle of Trafalgar the bosun who loaded Victory's mailbags forgot to include his own letter home to his wife. When word of this got to Nelson the mail ship was already a good way out, but the admiral called it back to pass the one letter, remarking that the bosun might well fall in battle the next day. These small concrete actions won his sailors' love in a way that no amount of grand speeches and posturing could ever have.

Insofar as control can be extended to social systems and command adapted in the face of nonconformant entitics, the key ingredient that can enable success is communication, and it is to this last member of our happy family we now turn.

## COMMUNICATION

Systems have structure, and a good way for this structure to be explored is by using the two families of *boundary, interior* and *exterior*, and *parts, relationships*, and *wholes*.

In addition to possessing structure, systems *do* things—they function. By putting the structure to work, systems also exhibit dynamics. Ideally, the dynamics of a system do not break its structure, and the structure of a system enables the dynamics to fulfill the system's function beautifully. For natural systems, this appears to be so, and these set the benchmark, providing exemplars for those that man designs and builds. The two families of *inputs, outputs*, and *transformations*, and *command, control*, and *communication* afford useful constructs for exploring a system's dynamics. Structure and dynamics, however, are not the whole of the system story.

A system undergoes change. Structures must necessarily change as the environment in which systems participate changes, and as the dynamics to which systems are subjected undergo change. A system therefore evolves, adapts, and learns. Whether this is over successive generations, as in the case of nature's species, or via successive revisions and innovations, as in the case of man-made systems, all systems undergo this inevitable evolution; in some cases, it's revolution—socially via political upheavals and technologically through disruptive innovation.

A system's dynamics exhibit flow, feedback, and fostering. This latter term is a description for system evolution. By fostering, we mean a system's ability to adapt its structure and update its dynamics in order to continue to function and so fulfill its purpose in widely differing environments that are uncertain, uncontrollable, unpredictable, and perhaps even unknowable.

One system construct we emphasize that points to how a system copes with this evolutionary demand is *communication*. Communication is the lubrication that enables a system to fulfill its function regardless of how its structure and dynamics necessarily change as its environment changes. Communication is a two-way street; it can exist within the interior of a system, for example, between its parts and between a system's interior and exterior, this latter case signifying communication between a system and its environment.

Communication of course is a field entirely of its own, and in our modern world, spans professionalisms that include the technological, the natural, the sociopolitical, and the cultural. The term is inevitably interpreted in many different ways, depending on which specialism is using it. As systems professionals we are interested in "specialism in breadth," and intend the meaning of communication to provide links and bridges across the disciplines bridging the divisions that spring up between them. We choose to make *communication* part of the systems language in a way that will foster an eclectic community of professionals, inspire systemic thinking, and, we hope, reduce the likelihood of systemic failure. We illustrate this by first describing two examples of communications systems, one from the natural world and the other from the sociopolitical-technological world, which you will find are comparable using *communication* as a key system concept.

## Ants

Ants are social insects. They form colonies that range in size from a few dozen predatory individuals living in small natural cavities to highly organized colonies that may occupy large territories and consist of millions of individuals. These larger colonies consist mostly of sterile wingless females forming castes of "workers," "soldiers," or other specialized groups. Nearly all ant colonies also have some fertile males called "drones," and one or more fertile females called "queens." The colonies are sometimes described as superorganisms because the ants appear to operate as a unified entity, collectively working together to support the colony. What enables a myriad of relatively unintelligent creatures to act as a smart unit?

Ants communicate with each other using pheromones. These chemical signals are more developed in ants than in other hymenopteran groups. Like other insects, ants perceive smells via long, thin, mobile antennae. The paired antennae provide information about the direction and intensity of scents. Since most ants live on the ground, they use the soil surface to leave pheromone trails that can be followed by other ants. In species that forage in groups, a forager that finds food marks a trail on the way back to the colony; this trail is followed by other ants, which then reinforce the trail when they head back with food to the colony. When the food source is exhausted, no new trails are marked by returning ants and the scent slowly dissipates. This behavior helps ants deal with changes in their environment. For instance, when an established path to a food source is blocked by an obstacle, the foragers leave the path to explore new routes. If an ant is successful, it leaves a new trail marking the shortest route on its return. Successful trails are followed by more ants, reinforcing better routes and gradually finding the best path.

Ants use pheromones for more than just making trails. A crushed ant emits an alarm pheromone that sends nearby ants into an attack frenzy and attracts more ants from further away. Several ant species even use "propaganda pheromones" to confuse enemy ants and make them fight among themselves. Pheromones are produced by a wide range of structures, including Dufour's glands, poison glands, and glands on the hindgut, pygidium, rectum, sternum, and hind tibia. Pheromones are also exchanged, mixed with food, and passed by trophallaxis, transferring information within the colony. This allows other ants to detect which task group (e.g., foraging or nest maintenance) other colony members belong to. In ant species with queen castes, workers begin to raise new queens in the colony when the dominant queen stops producing a specific pheromone. Some ants produce sounds by stridulation, using the gaster segments and their mandibles. Sounds may be used to communicate with colony members or with other species.

So, returning to the question, "What enables a myriad relatively unintelligent creatures to behave as a smart unit?" The answer, in a word, is *communication.*

## Arabs

Let's consider an entirely different question, "What enables millions of people who have been oppressed and tyrannized for thousands of years, each individual being regarded as having negligible worth and of course little or no military capability, to organize themselves into an indefatigable force that throws off the yolk of tyranny and to instigate an agreeable form of governance for themselves?" Might the answer to this completely different question be the same? Let's see!

Mohammed Bouazizi was a vegetable seller, one of hundreds of desperate, downtrodden young men in Sidi Bouzid, many of them with university degrees but without work and therefore forced to spend their days loitering in the cafés lining the dusty streets of their impoverished town. Bouazizi, exhausted by pushing his cart around all day, was glad to come home each night grateful for the meager living he was able to scratch out. His dream was to save enough money to be able to rent or buy a pickup truck, not to cruise around in, but to take the strain of his adopted labors. Not much of a dream. Yet Bouazizi, by the sacrifice of his life, cast a vision that would ignite the rage of a nation, topple its dictator, and enflame a cascade of revolt across a vast Arab landscape.

On December 17, 2010, Bouazizi's livelihood was threatened when a policewoman confiscated his unlicensed vegetable cart and its goods. It wasn't the first time it had happened, but it would be the last. Not satisfied with accepting the 10-dinar fine that Bouazizi tried to pay ($7, the equivalent of a good day's earnings), the policewoman allegedly slapped the scrawny young man, spat in his face, and insulted his dead father. Humiliated and dejected, Bouazizi, the breadwinner for his family of eight, went to the provincial headquarters, hoping to complain to local municipality officials, but they refused to see him.

At 11:30 a.m., less than an hour after the confrontation with the policewoman and without telling his family, Bouazizi returned to the elegant double-story white building with arched azure shutters, poured fuel over himself and set himself on fire. He did not die right away, but lingered in the hospital until January 4, 2011. Such was the outrage over his ordeal that even President Zine el Abidine Ben Ali, Tunisia's dictator for 23 years, visited Bouazizi on December 28 to try to blunt a nation's anger. But the outcry could not be suppressed, and 10 days after Bouazizi died, Ben Ali's 23-year rule was ended. But that was not the end.

One month after Ben Ali fled into exile in Saudi Arabia, a second dictator, President Hosni Mubarak of Egypt, stepped down following an 18-day revolt by the young people of that ancient nation. A mere 8 months later, that most ruthless of dictators, Muammar Gaddafi, was captured and shot. This after a long and bloody struggle between the military might of Libya and the impromptu galvanization of that nation's people, who were undoubtedly inspired to continue the movement begun by the martyrdom of an unknown, unlicensed purveyor of vegetables. Is this the end?

Algeria, Bahrain, Iran, Jordan, Lebanon, Syria, and Yemen are all experiencing cascading effects in different ways. All have civil society leaders who have made explicit claims of inspiration from Egypt and Tunisia. Is this movement, this so-called Arab Spring, actually a system? And if it is, what makes it possible to be formed as a whole from such a large assortment of autonomous, diverse systems hitherto unconnected and conceivably never before interrelated? How might a large number of these heterogeneous systems, for example, nation-states, political movements, aggrieved underclasses, and tech-savvy activists, hang together, and for the specific purpose of bringing democracy to a part of the world that has never known it before?

If we ourselves were pressed to answer the question: "How can the Arab Spring be viewed as a system?", we would be obliged to point to the phenomenom of communication and in particular to the communication platforms, that is, the digital media and their associated social networks, as the glue.

We take great care at this point. Digital media in and of itself cannot explain the decision that individuals make to face with no defense tear gas and rubber bullets. It cannot explain how citizens who have been denied basic human rights from birth are willing to face death at the hands of merciless oppressors who hold all the military cards. It cannot explain these things unless we look beyond the technology and see digital media *as* social networks. When credit is accorded, in all sincerity and by those with firsthand experience, to Facebook, Twitter, and YouTube for making the difference in how a people must behave in order to bring about long overdue social reform, we are obliged to judge these communication platforms less as graphical user interfaces and more in terms of their ability to achieve social mobilization.

Images of friends and family being beaten by security services draw people into the streets. Increasingly, those images are delivered digitally, as wall posts, tweets, and pixilated YouTube videos hastily recorded by mobile phones. Whenever social upheaval is identified at any point on the planet, the West is there to capture and broadcast the news to a global audience, fully armed with modern digital media. But the phenomenon of the Arab Spring has less to do with connecting Fleet Street with Arab Street and more to do with making connections between Arab streets.

The digital storytelling by the average Tunisian is what spread across North Africa and the Middle East. Social networks inhabit ground zero where digital media is now rooted. The pheromone trails that manifest the communication system of an ant colony are laid by the glands and detected by the antennae of individual ants. Likewise, news is made and read by individual human beings, some of whom self-immolate and others who bury their dead and thereafter live lives to honor that sacrifice by being willing to follow that blazed trail. Relating stories about shared grievances and a shared sense of desperation became much of the content communicated over these networks. The cascade effect wasn't simply that shared grievances spread from Tunis to Cairo, though. Instead, it was the inspiring story of success—the overthrow of Ben Ali—that spilled over networks of family and friends that stretch from Morocco to Jordan.

The content of this communication is personal and not ideological. In most social upheavals, there is an ideologically driven opposition that topples a dictator from another part of the political spectrum. Radical socialists, left-leaning union leaders, or a Marxist army from the countryside would lead a popular revolt. Or religious conservatives or right-wing generals would lead a coup. But the Arab Spring appears to be largely leaderless and without ideological labels. Political parties and religious fundamentalists are not the organizing principals. Instead, the upheaval is largely self-organizing, and this, by virtue of the communication to which we have alluded, is what chiefly makes it a system. Once the overthrow has occurred, the question arises as to what happens to this system. Can it be as effective at rest as it clearly had been in action? Though in many Arab countries there are now new leaders, new governance structures, and new hopes, there remains a considerable amount of skepticism and malcontentedness among the ordinary people. Might this continue to be the fuel to continue the struggle? Whether the Arab Spring system remains in place or not, the communication afforded by digital media and social networks is as potent as ever and permanently available.

Over the last decade, information and communication technologies have had consistent roles in the narrative for social mobilization. As one activist successfully tweeted about why digital media was so important to the organization of political unrest, "We use Facebook to schedule the protests, Twitter to coordinate, and YouTube to tell the world." Protesters openly acknowledge the role of digital media as a fundamental infrastructure for their work. Digital media allows foreign governments and diaspora communities to support local democratic movements through information, electronic financial transfers, offshore logistics, and moral encouragement. It further supports the organization of radical student movements to use unconventional protest tactics at sensitive moments for unpopular regimes. It unites opposition movements through social networking applications, shared media portals for creating and distributing digital content, and online forums for debating political strategy and public policy options. It attracts international news media attention and diplomatic pressure through digital content, such as photos taken "on the ground" by citizens or leaked videos and documents to foreign journalists, or through diplomats raising flags over human rights abuses and political corruption. Finally, it provides a transport mechanism for mobilization strategies from one country to another, sharing stories of success and failure, and building a sense of transnational grievance with national solutions.

Digital media and associated social networks clearly possess a higher level of sophistication than the humble ants' pheromone trails. But then ants don't get themselves enslaved as people seem so often to do. They plan, work, and cooperate without the aid of any commanders. Thankfully, communication, as a concept, is agnostic to the system it can so capably hold together.

## THE FUTURE IS NOT WHAT IT USED TO BE

The previous four chapters showed us that in regard to any system of interest, we discover two salient characteristics: *structure* and *process*. Structure gives us an idea of what the system is, and process gives us a picture of what the system does, or more correctly, how the structure behaves in a dynamic fashion.

Chapters 8 and 9 introduced two families of triples that enable us to explore system structure. Chapters 10 and 11 introduced two more families that help us better understand how a system's structure operates dynamically, which is to say which process takes place that shows us a system's behavior. But there is, as you might expect, more to a system than these two characteristics of structure and process. And this is explained by a single word: change.

All systems undergo change. The exterior of an environment undergoes continuous change, and so a system's interior must change, which is to say its structure and process must change if the system is to cope. Either that or the system's boundary must be impermeable to the impacts of change, but this measure might very well prove pejorative to a system's well-being. By the same token, the managers of a system are under constant pressure to improve a system's performance, and this they often achieve by making changes to a system's interior and to its boundary.

The overarching theme of the last three chapters in Journey II relates to the implications and consequences of change for any given system of interest. In particular, we want to know what the important ideas are that we should associate with a system of interest that we know will help it to respond to changes and enable it to survive and ideally prosper as the inevitable force of change impacts a system in various ways.

We will introduce three more families of triples. The first of these is *structure, process, and function*. Notice that what we have done here is to embed the ideas of Chapters 9 (structure) and 10 (process) in a higher level fashion. By doing so, we are pointing to the fact that these two (structure and process) work together to serve a higher-order purpose, one that enables a system to operate in an environment of perpetual change. In order words, regardless of changes that may affect a system's interior and exterior (and indeed boundary), a system must continue to perform its designated function. This feature of going to a higher level is fundamental to systemic thinking.

The second family is *variety, parsimony, and harmony*. This set of ideas will help us to understand such matters as beauty, economy, elegance, and constraint, all of which are needful considerations given that a system must bear up under the force of change. The final triple is *openness, hierarchy, and emergence*. All systems must have some degree of openness in order to survive in the environment in which they are located. The nature of openness naturally leads to hierarchical organization, which is a common and powerful pattern for both structure and process. Emergence is perhaps the chief of all system ideas. It is what distinguishes the whole from the summation of parts and relationships, what distinguishes function from the summation of structure and process, and what distinguishes communication from the summation of command and control. These three families will then complete our package of 21 system concepts. In total, these provide all those with an interest in better understanding the nature and purpose of systems with an enviably comprehensive arsenal of ideas.

# CHAPTER 12

# STRUCTURE, PROCESS, AND FUNCTION

When we first looked at the human heart in Chapter 9, we were primarily concerned with its structure. This is what we said:

> Located between the lungs in the middle of the chest, behind and slightly to the left of the breastbone (sternum), the heart is surrounded by a double-layered membrane called the pericardium. The outer layer of the pericardium surrounds the roots of the heart's major blood vessels and is attached by ligaments to the spinal column, diaphragm, and other parts of the body. The inner layer of the pericardium is attached to the heart muscle. A coating of fluid separates the two layers of membrane, letting the heart move as it beats, yet still be attached to the body.
>
> The heart has four chambers. The upper chambers are called the left and right atria, and the lower chambers are called the left and right ventricles. A wall of muscle called the septum separates the left and right atria and the left and right ventricles. The left ventricle is the largest and strongest chamber. The left ventricle's chamber walls

*Systemic Thinking: Building Maps for Worlds of Systems*, First Edition.
John Boardman and Brian Sauser.
© 2013 John Wiley & Sons, Inc., Published 2013 by John Wiley & Sons, Inc.

are only about half an inch thick, but they have enough force to push blood through the aortic valve and into the body.

Four types of valves regulate blood flow through the heart: the tricuspid valve regulates blood flow between the right atrium and right ventricle; the pulmonary valve controls blood flow from the right ventricle into the pulmonary arteries, which carry blood to the lungs to pick up oxygen; the mitral valve lets oxygen-rich blood from the lungs pass from the left atrium into the left ventricle; and the aortic valve opens the way for the oxygen-rich blood to pass from the left ventricle into the aorta, the body's largest artery, where it is delivered to the rest of the body.

So in essence, the heart is a bodily organ attached to the sternum consisting of four chambers and four valves regulating the flow of blood between the chambers and between the heart and the rest of the body. Clearly, we could look deeper into this structure, but right now that's not our direction of travel.

We also looked at the regulation of blood flow in Chapter 9 and more deeply into this process in Chapter 10. Here is what we said, briefly:

Electrical impulses from the myocardium cause the heart to contract. This electrical signal begins in the sinoatrial (SA) node, located at the top of the right atrium. The SA node is sometimes called the heart's "natural pacemaker." An electrical impulse from this natural pacemaker travels through the muscle fibers of the atria and ventricles, causing them to contract. . . . Blood is carried from the heart to the rest of the body through a complex network of arteries, arterioles, and capillaries and is returned to the heart through venules and veins.

A heartbeat is a two-part pumping action that takes about a second. As blood collects in the upper chambers (the right and left atria), the heart's natural pacemaker (the SA node) sends out an electrical signal that causes the atria to contract. This contraction pushes blood through the tricuspid and mitral valves into the resting lower chambers (the right and left ventricles). This part of the two-part pumping phase (the longer of the two) is called diastole. The second part of the pumping phase begins when the ventricles are full of blood. The electrical signals from the SA node travel along a pathway of cells to the ventricles, causing them to contract. This is called systole. As the tricuspid and mitral valves shut tight to prevent a backflow of blood, the pulmonary and aortic valves are pushed open. While blood is pushed from the right ventricle into the lungs to pick up oxygen, oxygen-rich blood flows from the left ventricle to the heart and other parts of the body. After blood moves into the pulmonary artery and the aorta, the ventricles relax, and the pulmonary and aortic valves close. The lower pressure in the ventricles causes the tricuspid and mitral valves to open, and the cycle begins again.

So what we are finding now is that there is a process involving the structural elements of the heart working dynamically together—structure and process—to realize function. And what is the function of the human heart? Simply this: to pump blood! If there was a better way to do this, one could find a better structure, or a better process, or both. But whatever "improvements" were found, one thing is essential: the structure and the process must work together and in so doing they must realize the declared function. It might sound obvious, but it is nevertheless profound. And structure process and function are a cornerstone set of ideas that makes something worthy of the name *system*.

We use this little family now to ask: "Why this function? What purpose does it serve? If the human heart is a system, what is the world of systems to which it belongs?" Clearly, we might ask these very same questions about these other systems in that world. Our direction of travel using this triple then is onward and upward, or at least outward. It is the beginning of navigating worlds of systems.

We urge you to do this for your own benefit relative to the human heart and to the wider anatomy and physiology of its world of systems. But our interest now turns to illustrating the virtue of this triple by exploring two rather different systems, both concerning sexual activity, with one being about the imperative of procreation and the other relating to the spread of HIV.

## FLASH (OF BRILLIANCE)

Imagine a 200-yd stretch of forest bordering a river in Papua New Guinea. The trees are 40 ft tall. The scene, while verdant and panoramic, is nothing extraordinary. A firefly decks each leaf, but you see none. Night falls. Speckles of light dapple the stretch and interest awakens. Soon there are clusters of blinking lights as near neighbors get accustomed to their fellow flashers. The scene crescendos in a series of single solid flashes, about twice per second, along the entire stretch of river. Millions of fireflies have synchronized themselves into an orchestra of light in the darkness. It is a sight to rival anything that Walt Disney pulls off at Epcot. The scientific term is terrestrial bioluminescence (Akilesh 2000). Fancy title, but what's it all about?

The flashing is the means by which males attract females prior to mating. It makes sense to produce a series of single blinding flashes. That will get the females' attention. Going it alone is risky. But can a forest of fireflies produce the collective consciousness to synchronize in order to maximize mating potential? Who has that idea? Some of them or all? And how do they share that notion? Is there a conductor for this orchestra? The ant, we are told, has no commander. Are fireflies somehow a more intelligent species, so that a leader emerges from their uniform ranks? And if so, is it possible for millions of fireflies to notice a single leader and be smart enough to subordinate themselves to this single command? Let's think about this systemically.

Using our triple, we state that the function we are faced with is "to create a sustained series of brilliant flashes of light for the sole purpose of reducing confusion in the female population and so attract its members to the light so that sexual activity can take place and thereby safeguard the continuation of the species." That is the function we wish to reproduce. So what's the system? What is its structure and what is its process? And how do these two work together to realize that very function?

If there is a system, then the male fireflies must surely be crucial parts. What are their relationships? Let's suppose that each firefly is influential in some way on some others and each is influenceable. If there were 10,000 fireflies, then the total number of communication paths between them, whereby any firefly could influence and be influenced by any other, would be 50 million. If there were 100,000 fireflies, then the total of pathways would rise to 5 billion. It's just not possible to imagine the humble firefly being capable of paying attention to 100,000 (or even 10,000) chat lines in the context of 5 billion (or 50 million) such chats. It's not possible.

This would lead us to conclude that influence must be confined to near neighbors. By influence, we now mean responsiveness to the single flash of another firefly, and by responsive we mean making an attempt to synchronize. So that in a connection between any two fireflies, there is the possibility that they will flash at the same time (more or less). A neighborhood of 20 fireflies would limit the maximum number of pathways to 190. In the case of any firefly being connected to, say, 10 others, the neighborhood of 20 would be a tight-knit group, which appears reasonable. There would have to be 500 neighborhoods to make up a total population of 10,000, which is also conceivable, but would this structure account for a global synchronization to produce a flash of brilliance?

We now have a structure with 500 subsystems, each having 20 parts and a total of about 100,000 connections (reduced from an original maximum of 50 million, or 0.2%). However, these "subsystems" are not connected, so no flash of brilliance can be expected. What is missing that might be possible?

A rare few fireflies might feel the influence of a firefly or two at a longer distance, that is, between distant neighborhoods. A few fireflies might have a particularly brilliant flash, and so be visible to others far away, or a few genetic oddballs might respond more to a fainter flashes than to bright ones. In either case, some fireflies make it possible for long-distance links to exist between these neighborhoods. This argument begins to present the opportunity of a small-world architecture whereby any male firefly could potentially be influenced by any other via, say, three connections. It turns out that only a very tiny sprinkling of such distant influences would need to be in place to achieve this togetherness. If there happened to exist around 25,000 new inter-neighborhood connections (this being 0.05% of the original 50 million for a fully connected structure or, viewed another way, an additional 4% of the grand total of intra-neighborhood connections), a small world would result.

With that rethought structure, the process needed to achieve global synchronization is relatively simple. Each firefly is sensitive to the flash of a neighbor (with this term now including fireflies in remote neighborhoods). That sensitivity results in synchronization of flashes between the neighbors. This produces a brighter light that will be noticed by other neighbors, who will then also look to synchronize. And that cascading effect is how the synchronization spreads and how the brilliance of the flash increases, resulting in the desired function. For your further edification and enjoyment, we encourage you to take a look at Steven Strogatz on TED.com (Strogatz 2008).

## DON'T STOP ME NOW!

Freddie Mercury was undoubtedly a talented artist. A scintillating performer who fronted the rock band Queen, when he passed, the remaining three members of the group disbanded. The band's records continue to sell, and *We Will Rock You*, the jukebox musical that incorporates hits from Queen, remains a worldwide hit even today, having opened some 10 years after Freddie's passing in 1991.

The man has gone, his legacy remains, but his actual death was a landmark. Mercury was the first major rock star to die of AIDS, and his death represented a very important event in the disease's history. Awareness of the disease spread, and millions were raised for AIDS research, many via concerts organized by the Mercury Phoenix Trust founded by Queen. Some argue that more could have been done had Mercury been more open about his illness, which was first diagnosed in 1987. An incomparable extrovert on stage, Mercury was an exceedingly shy and private person off it. It's hardly unsurprising, therefore, that he should choose not to speak of his condition or situation, even though this may have been disadvantageous in the broader social sense. That really only adds to the tragedy. HIV/AIDS is a major health problem in many parts of the world and is considered a pandemic—a disease outbreak that is present over a large area and is actively spreading. As of 2010, approximately 34 million people have HIV globally. Of these, approximately 15.9 million are women and 3.4 million are younger than 15 years old. HIV/AIDS resulted in about 1.8 million deaths in 2010, down from 3.1 million in 2001. Since AIDS was first recognized in 1981, it has led to nearly 30 million deaths, as of 2009 (statistics published by UNAIDS, WHO, and UNICEF in November 2011 [UNAIDS 2011]). It's as if the disease itself is rehearsing the words of the Mercury hit "Don't Stop Me Now" ("'cos I'm having such a good time") . . .

I'm a rocket ship on my way to Mars
On a collision course
I am a satellite
I'm out of control
I'm a sex machine ready to reload
Like an atom bomb about to oh oh oh oh oh explode!

How do you stop this sex machine, defuse this atom bomb, get this disease under control, and prevent this rocket ship from colliding (with millions of others)? Some of the AIDS research is directed toward a vaccine. No cure has yet been found, though some treatments slow the progression of the disease. A complementary approach is that of awareness programs, which range from encouraging sexual abstinence to sexual education classes that make people, especially the young, aware of the disease, how it is transmitted, what the risks are of various practices, and what steps can be taken to reduce or eliminate these risks. It is this social aspect of the disease that we reflect upon briefly now using some systemic ideas.

HIV/AIDS is transmitted by three main routes: sexual contact, exposure to infected body fluids or tissues, and from mother to child during pregnancy, delivery, or breastfeeding (known as vertical transmission). We will confine our attention for now with the first of these modes only.

Let us suppose that there is in place a system of human sexual activity and that its function is to bring satisfaction through sexual intercourse between consenting individuals. We want to know what is the structure and process of this system, one reason being to understand how the HIV/AIDS disease has managed to infiltrate this system and somehow spread itself very widely across its structure leveraging its process. If we can get some answers to the structure and process of this system, especially answers that support the evidence of the dramatic spread of the HIV/AIDS disease, we might then be able to posit some strategies for defending against this transmission. These would be measures to be taken that prevent its ultimate collapse, meaning the deaths of those individuals who make up the system.

Let's picture the human individuals in this system as nodes and a bilateral sexual relationship between two individuals as a link connecting these two nodes. Some individuals will abstain from sexual activity altogether. They are not part of this system because they have no connection. Others will have a single partner and remain faithful to the other for life. If this is reciprocated, these two will not be part of the system since although they are connected to one another, they have no connections to the system. They are their own system. The system of interest includes a large number of individuals who have multiple sexual partners and (unfortunately) those individuals faithful to one partner who has other sexual partners. We want to know, "What does this system look like? What is its structure? Is it for example like the structure we discovered that underlies the syncopated firefly flashers?"

That structure is known as an egalitarian network. It is so called because the network consists of a huge number of relatively small subnetworks (of equals). Subnetworks are massively tight-knit, with almost all of their elements connected directly to the majority of the other members. A few of these elements also have direct remote connections to members of other subnetworks. These connections are known as weak links because they are isolated and probably infrequently exercised. Their existence, however, curiously explains the strength of the large network, the system as a whole, or the world of systems (subnetworks).

Does our human sexual activity system resemble an egalitarian network? If it does, then these weak links are the culprits for spreading the disease. They would be impossibly difficult to spot, which might explain the intractable problem of inhibiting transmission of the disease. A more likely explanation, however, is that the structure of our system is that of an aristocratic network.

In this architectural pattern, there are a few super hubs, individuals who have an exceedingly large number of sexual partners. Further down the scale, there are many more individuals with very much fewer partners, and at the "bottom of the scale," as it were, are the vast majority of individuals who have a handful of partners, including those who have only one partner for life. This scale is known as the power law, and it has been employed to depict a wide range of systems, including river drainage, the Internet, and the World Wide Web, among others. What is the process that drives this architectural pattern, that is, which gives rise to the super hubs or significant few individuals? The answer lies in the forces of growth and preferential attachment. Growth simply means the addition of nodes and of links. In other words, more people enter the system over time (some leave, of course), and more relationships spring up (for a variety of reasons that you can no doubt imagine). By preferential attachment, we mean that it is considered more desirable to have a relationship with an experienced other, in other words, to have gone out with a super stud as opposed to someone inexperienced (this may or may not be true). Forces are certainly at work that drive the power of the stud. With success at gaining new partners comes an acquired skill to gain yet more. With more partners gained comes the need to practice that skill more extensively in order to keep up a good image. With that motivation comes more skill and more partners. It is a cycle that Peter Senge would call a reinforcing loop and characterize by a snowball rolling down the side of a mountain, potentially producing an avalanche. This is the sex machine that is ever ready to reload, who insists, "Don't stop me now."

The danger of this aristocratic network is that howsoever an infection is introduced into the network (and it may be via an individual at the bottom of the scale), once it gets to the super hub, it inevitably and irresistibly spreads. As long as the network remains as it is, aristocratic, then everyone will be infected. There is nothing that will stop it. Such networks don't tip; by their nature they already have. The only prevention is the rearchitecting of the network; it has to be broken up. Strangely, the super hubs are not just the dominant feature of the network; they are also its vulnerability. Take them out and the network splinters. This is the vulnerability of the Internet, which for good reason therefore provides unassailable protection and security to its super hubs. However, taking out the super hubs in the sexual activity network is nontrivial. They have first to be identified, which is no easy matter, given the privacy that individuals fiercely protect. And having identified them, they have to become aware of their status in the network and make very certain they themselves do not contract the disease. As challenging as this task is, it does with reasonable certainty deal with the problem. Medical solutions will continue to be sought, as indeed is right. However, social solutions are also worth pursuing, and in this regard, systemic ideas are there to help.

# CHAPTER 13

# VARIETY, PARSIMONY, AND HARMONY

## UNDER PRESSURE

*The Remains of the Day* is a novel by Kazuo Ishiguro. It tells the story of a loyal and dignified butler, James Stevens, and his unstintingly steadfast service to his master, Lord Darlington, in the years leading up to World War II. The book was made into a movie by the magnificent team of James Ivory and Ishmael Merchant, with Anthony Hopkins playing Mr. Stevens and James Fox playing Darlington. Lord Darlington's sympathies for Germany, and most dangerously for the emerging Nazi party that sought restoration of their nation's prominence following its brutal subjugation after World War I, contributed inevitably to his downfall. His personal decline was a poignant illustration of the gradual demise of the aristocracy in Great Britain. This erosion of power among a ruling elite was accelerated by the Parliament Act of 1911 and formed a key part of the context for both book and the film.

*Systemic Thinking: Building Maps for Worlds of Systems*, First Edition.
John Boardman and Brian Sauser.
© 2013 John Wiley & Sons, Inc., Published 2013 by John Wiley & Sons, Inc.

In one scene, Darlington and his friends are in the drawing room enjoying their after-dinner scotch served by Stevens. The conversation turns to the matter of accounting for the views of the great unwashed public in the strategic affairs that might affect the nation's decision making. One argues for consultation, valuing the opinions of the ordinary man on the street and his right to express them. Another, Spencer, demurs. The latter seeks to prove his point by interrogating the unsuspecting Stevens, with the initial approval of Darlington. The dialog runs along these lines (Ivory 1993):

*DARLINGTON:* Oh, Stevens; my friend would like to ask you a question.

*STEVENS:* Sir?

*SPENCER:* My good man, do you suppose that the debt situation regarding America is a significant factor in the present low levels of trade, or is this a red herring, and that the abandonment of the gold standard is at the root of the problem?

*Stevens stands motionless:* silent and bewildered, hesitant yet dignified. He replies that he is unable to be of assistance in the matter.

*SPENCER:* What a pity. Let me ask you then, do you think that the currency problem in Europe would be alleviated by an arms agreement between the French and the Bolsheviks?

Stevens must wonder what is happening to him. He stands in service to his master, who is now clearly showing signs of discomfort, wishing to put an end to the torture. Holding to his dignity while remaining ever loyal to his lord, Stevens once more replies that he is unable to be of assistance in the matter. Is this the end? It is not.

*SPENCER:* Well perhaps you can help me on this matter. Do you share our opinion that Monsieur Daladier's recent speech on the situation in North Africa is merely a ruse to scupper the nationalist fringe of his own domestic party?

No end in sight for Stevens. No change of demeanor. No variation in response.

*STEVENS:*  I am sorry sir, but I'm afraid that I am unable to be of
assistance in any of these matters.

*DARLINGTON:*  That will be all, Stevens.

*STEVENS:*  My lord. Sir. *(looking at Spencer)*

The final word goes to the smugly satisfied torturer who feels that
he has more than proved his point. He scoffingly asserts that it is
foolishness to suppose that Stevens and the millions like him can
know anything of value that relates to the great affairs of State and
that to give them any opportunity to become involved in such decision making would be like inviting a committee of the Mothers'
Union to organize a war campaign.

The system of knowledge possessed by Mr. Stevens was not
adequate to the external scrutiny of one of his fellow citizens, argu-
ably one of his superiors. Stevens found himself in a hostile environment, and though he did not crumble, he was nonetheless
stupefied. It was a situation in which he could not survive let alone
prosper. He had to be excused. The variety of his resources was not
adequate to the dimensions of the environment in which he was
(unfairly) placed. We see this time and again. In nearly every case,
the system does not survive. Only in those instances where a system
can adapt, by somehow finding that extra variety of resources can
that type of system expect to endure and by so doing pass on its
redeveloped resource base to successor generations. Interestingly,
Stevens, because of his dignified manner and unswerving loyalty to
his lord, was able to "degrade gracefully at the boundary." While
this did not win him any arguments or the subsequent approbation
of his aristocratic interrogator, it did very likely win him some
admirers at least from the readership or moviegoers. Sometimes, it
is possible to deal with external pressures not by generating an
equal and opposite pressure but by a change of focus, by a tender
deflection, one that replaces a traditional respect for knowledge
with an emerging admiration for gentleness.

## LET YOUR WORDS BE FEW

Spartacus was born a slave. He lived his entire life as one, if living
is indeed an apt description for such an existence. Death was the

only available exit door from slavery. To a slave, death means freedom, which in that sense made Spartacus fearless. He was also strong and resolute. It was a powerful combination that did not go unnoticed by Batiatus, the owner of a gladiator school, which was why he was happy to part with a few denarii to purchase Spartacus from his Roman captors. And with that exchange begins a new phase of slavery where once again the only freedom is via death in the ring at the hands of a fellow gladiator in a contest put on for the pleasure of the senseless wealthy elite of the Roman empire.

Spartacus learns the skills that make him one of the best, if not the best of all gladiators. In one contest, witnessed by Crassus, an immensely wealthy citizen, Senator, and commander of a private army to rival that of Rome, Spartacus defeats a fellow gladiator but refuses to kill him, which is the wish of Crassus' friends for whom the contest has been especially laid on. This act of mercy marks Spartacus the man and identifies him as a potential leader of a future slave army. In an episode of high drama, the gladiator whom Spartacus had defeated and whose life was spared chooses to lose it in a vain attempt to kill Crassus.

On one particular evening, Batiatus benevolently permits each gladiator to receive a female escort from among the servants owned by Batiatus. This is intended as a sort of morale-boosting promotion for the athletes designed to make future contests more combative and worth winning. Over a long period of time, it is actually possible for a gladiator to secure his freedom and become one of the trainers at the school, as was the case for Marcellus, the brutal and remorseless chief trainer who watches Spartacus like a hawk and would willingly end his life rather than allowing him to become a threat. So the death of a fellow gladiator might be the way to end slavery without losing one's own life.

Spartacus is visited by Varinia in his private cell. She is incomparably beautiful. Someone worth living for, and that would be an apt description. Spartacus is captivated. Motionless at first, stunned by her beauty, he moves slowly and tenderly around her, taking in this object of desire to whom he would willingly give all his love were he only to know how. He tells her that he's never had a woman before. It's evidently not her first time. She moves into the shadows and begins to disrobe, knowing to ignore any reluctance she has about enforced pleasure, doing her duty as the slave that she is. No freedom to love even though love is all around.

Except that above there is no love, only a base voyeurism. In the ceiling of the cell is a barred opening, and from this bird's-eye view leer Batiatus and Marcellus. Spartacus is horrified and outraged and does all he can physically to show his contempt for these unwanted evil onlookers. The two scoff. Spartacus screams that he is not an animal. He repeats the line just as insistently but more gently, and now in a manner that seeks Varinia's understanding. He plaintively cries, "I'm not an animal."

She looks at him with remarkable sympathy and understanding. She waits to reply, an age seemingly. Perhaps she'll say nothing. Why should she? Then utters pithily, "Neither am I." A simple line. Just three words. They say it all.

The door of the cell opens. Batiatus and Marcellus enter. Batiatus tells Spartacus he is disappointed with him. Varinia is escorted from the cell with Batiatus now rubbing salt in the wound: "She'll sleep with the Spaniard tonight!"

Spartacus is left alone, with just his thoughts, of what might have been and probably never will be again. The door slams shut. Is this the end, even of existence? Some days later, the gladiators are seated waiting for lunch to be served. Varinia is one of the serving girls. She tracks down the line of men filling their individual basins with that day's gruel from the large serving bowl she carries. Speaking is not allowed. All are watched. A tongue will be ripped out if a word is spoken. Silence reigns but Spartacus' mind is in overdrive as he sees Varinia approach. He feels compelled to speak. He must speak. But what can he say? What dare he whisper? What's on his mind?

We have described this scene on many occasions to our classes of Master's students. It's not simply that movies are a major point of communication between professor and student, though indeed they are. And it's not as though the 1960 movie with a stellar cast that includes Kirk Douglas (Spartacus), Peter Ustinov (Batiatus), Lawrence Olivier (Crassus), and Jean Simmons (Varinia) is all that well known today or much appreciated. It is done because having relayed this part of the movie we challenge our students to write the missing line—*using exactly four words*. It is an exercise in parsimony and it is not easily executed, as so many students can attest.

Parsimony is not just meanness or minimally extreme. It is enriched meanness. It is an extremum in economy that also takes account of the extremum of variety. The four words that Spartacus spoke must capture everything that has gone before. Yet they must also be the words of a simple, confused inarticulate slave whose only physical skills are to wield gladiatorial weapons and whose emotional intelligence is highly developed. We challenge you to find these four words. More important, use as many words as you care to take, and also explain the reason for your choice of these four words.

## EBONY AND IVORY

Stevens and Spartacus are complex systems. Although fictional characters, inventions of the human mind, they are no less complex for that. Each character is embedded in a social context and interacts with that context using peculiar and personal characteristics, social skills, beliefs, motivations, and so on. These have to be given them by their creators, but having thus been endowed, the characters must behave consistently though possibly unpredictably. The reader—or viewer—would be decidedly unhappy if there were no pattern or "fingerprints" to these characters.

Their inventors, in the case of Stevens it's Kazuo Ishiguro, and in the case of Spartacus, it's a combination of Dalton Trumbo (screenwriter) and Howard Fast (novelist), are also complex systems. Each of these is embedded in social contexts and to each is given capabilities, viewpoints, priorities, desires, and so on. There is no reason why their inventions should be any less complex than the real thing.

We employed each of these characters to illustrate a particular systemic idea. In the case of Stevens, this is *variety*, and in the case of Spartacus, it's *parsimony*. We could have switched these two around using different episodes from the stories to which they belong. Both characters come under pressure, and both are capable of exemplifying parsimony—in speech as well as action. Every one of us every day comes under some form of pressure. Likewise, each of us, from time to time, gets to exercise the gift of parsimony by being to the point, relevant, impactful, and efficient. There's nothing bad about coming under pressure; matters are not made worse by collapsing under pressure. The point is that environments for all complex systems can prove hostile, menacing, and fatal. This sounds bad. The good that it does is to provide the opportunity to learn and to adapt, if not for the individual who does not survive, then for his or her fellow man via the lessons learned.

This we know: that to survive and prosper in an environment, a system must possess requisite variety, this being the number of "degrees of freedom" that the environment has. For "degree of freedom," substitute strategies, choices, capabilities, resources, knowledge, and attitude, among many others. Requisite variety calls for a comprehensive portfolio of survival capabilities, at least sufficient to match the environmental challenge. Requisite variety is a maximizing force. But no system can be influenced by this force limitlessly. There isn't time, there isn't space. Googol is not a big enough number. What restrains requisite variety, allowing it to find an agreeable extremum, is parsimony.

There is something beautiful, elegant, and exquisite about parsimony. It's not, as we have said, meanness for the sake of meanness, but rather essence: the essential minimal result of much observation, analysis, contemplation, processing, and searching for exemplars. It is a matter of keeping things simple, of making them as simple as possible—and no simpler. It is the proper application of Ockham's razor. For electrical engineers, it is the search for using fewer amps, fewer watts, and less mass, all while simultaneously searching for maximum functionality of the device being designed. For project managers, it is the search for less time and reduced costs while delivering a product or service that not only satisfies the customer but delights him or her.

These two need each other. On their own, they are a menace to the system. Balancing each other out, they produce a system that will cope with change, learn from mistakes, do better next time, and strive for excellence. They are the black and the white keys on a keyboard. Both are needed for harmony. Working together, they entitle the object to be properly called a system, one that will crescendo in maturity.

# CHAPTER 14

# OPENNESS, HIERARCHY, AND EMERGENCE

And so we come to the final family of triples, the last piece of the scaffolding, the remaining element of our framework of systemic ideas. Did we save the best until last?

## NOW I'M HERE

To some degree or other, all systems must be open. An entirely closed system could not take in any new information nor could it take on board any new energy. Such a system would have to have all the information it needed to survive and prosper *ab initio*, or be omniscient! If it kept to docile environments, that might be okay but this is not something that the system could guarantee, since it would be unable to exercise any kind of control over the environment on account of the fact that it would have no communication

*Systemic Thinking: Building Maps for Worlds of Systems*, First Edition.
John Boardman and Brian Sauser.
© 2013 John Wiley & Sons, Inc., Published 2013 by John Wiley & Sons, Inc.

with its environment. On the energy front, unless the system had a means of renewable energy, then it would presumably expend its original supply, and once this had totally expired, it too would cease to be. Again, there's no problem here if this was what had always been planned—limited lifetime duration. Some systems have very specific missions involving fixed time horizons of existence and fixed degrees of freedom, relative to the knowledge required to accommodate the mission. Under these circumstances, it could indeed be highly closed. The advantage of tight closure is minimum or even zero risk of interference with its knowledge, energy supply, and mission details. But if for any reason the system encounters surprises in its trajectory toward its goal, then it will almost certainly be unable to deal with these since it will lack adaptive capability.

Being closed would also mean that the system could not eject waste. If this waste could be recycled and at the same time provide a source of new energy, that would be some form of compensation, and indeed highly closed systems necessarily take this feature into their initial design. But some information can be regarded as waste since it is no longer valid, becoming out of date or later proved to be erroneous. Such sources are a nuisance and can interfere with vital knowledge-processing mechanisms. Measures would need to be put in place to recycle this information, making better use of storage. However, this almost certainly requires an openness of the system to revised viewpoints and more up-to-date information.

The reality is that no system is entirely open or entirely closed. In the ideal case, the system would be open when necessary and closed otherwise. Knowing when to be in which state requires maturity and judgment, and these qualities speak to a higher level of openness. A study of openness requires us to think about the reasons why a system should be open (to some things and at certain times) and closed (otherwise).

At the first level, openness is all about transactions. An example of this is trading. One party has some goods and services and it is are prepared to sell them to another party for some amount of currency or trade them in exchange for other goods and services. In fact, currency itself can be traded and exchanged, as indeed can interest rates, which apply to contracts of exchange. The world of finance has developed trading to a level that many would find utterly bewildering, and yet the livelihoods and futures of many of us in terms of our pensions, for example, depend upon these mysterious trades.

At another level, openness is about relationships. Living organisms are in relationship with one another and with their environment, which includes both prey and predator. Relationships can be cooperative, competitive, or combative. Corporations have been compared to living organisms, and it is certainly the case that cooperation, competition, and adversarialism exist in various forms between corporate organizations. Not only between but within, all the way "down" to the level of the individual worker. What was it that Michael Corleone was taught by his father? "Keep your friends close and your enemies closer"! Openness to enemies sounds like an unusual circumstance; nevertheless, it is one the wise do well to heed.

Finally, openness is motivated by strategic intent. By this we mean a system's desire to survive and prosper regardless of environmental changes, forces of cooperation and competition among and between species, availability of information and degree of misinformation, and changes in context and culture. It is at this level that one learns best to know when and how to be open or closed. Camouflage is perhaps a beautiful example of strategy. A system is wide open to the environment and therefore able to roam, explore, and gather supplies and information, and yet it is closed to predators (and competitors) since it blends into the background, proceeding largely unobserved as it goes about its mission.

## COME TOGETHER

As systems open themselves up to one another, via trading in the first instance, perhaps, they develop their relationships and become trusting of one another. Learning how to trust is an important feature of a system's existence. Once that trust reaches a certain level, it is then quite common for these systems to be joined together as one.

In the case of organizations, some of these unions reflect a genuine bilateralism. For example, when the Northrop Corporation joined forces with the Grumman Corporation, it formed Northrop Grumman (NGC). When you do business with NGC, it is not difficult to tell which employees formerly belonged to which constituent. Ideally, the union was synergistic, leveraging the skills and resources of each constituent for the common good, in addition to achieving the economies of scale that are so often declared to shareholders as being profitable.

Other kinds of union are more of the acquisition type than of merger (of equals). So, for example, Microsoft acquires Skype. The latter company then no longer exists, although the name lives on as a brand and identifier of, in particular, Voice over Internet Protocol (VoIP) capability. Microsoft has plans for that capability, and these are gradually unfolding for public view, with Skype being rolled into Outlook, creating an integrated platform for messaging, video, and social networking. The question posed of this type of union is "Will the culture, the modus operandi, and the very driving force behind the acquired company be allowed to continue now that it has become part of Microsoft?" For it to be any less could destroy its value and prove an expensive and nugatory effort by the acquirer. But for it to go unchanged might mean that Microsoft itself has no corporate culture but is instead a patchwork quilt made up of the idiosyncratic cultures of all such acquisitions. That makes for an enormous challenge in terms of integration, which is the byword for that industry in the modern era. Ideally, one would wish to preserve individuality while promoting unity. One form of structure that goes some way to achieving this is *hierarchy*, and this is both naturally occurring and an agreeable cousin to openness. Let's look at some prime examples.

The basic building block in all electronic devices is the transistor. This is a semiconductor device used to amplify and to switch electronic signals and electrical power. These devices first appeared in discrete form, being intended to replace vacuum tubes, which performed essentially the same functions but requiring more power, creating more heat, and taking up far greater space.

The discrete transistor ushered in the era of miniaturization. Today, billions of transistors are hidden away in sophisticated integrated circuits (ICs) that can, as a result of the complicated circuitry of transistors that they contain, perform a multiplicity of functions, many of which are vital to the computing devices we have today. The integration of basic building blocks into higher-level devices is a form of hierarchy that brings unparalleled levels of organization and logic to our world.

ICs as discrete components are themselves assembled into circuitry that then provides us with mobile phones, tablets, laptops, and a vast array of labor-saving and smart devices in homes, offices, and factories around the world. Using the Internet and the variety of virtual computing media, these components can be brought together to provide the sophisticated systems for sensing, monitoring, commanding, controlling, computing, communicating, sharing, and integrating data, information, and services that we now consider vital to our global economy and international community.

This pattern of integrating basic components that lead to new devices and systems that can then be integrated into higher-order systems is what we know as hierarchy. This is a powerful form of organization that we all recognize. It relies on the ability to integrate individual elements and by so doing produce new higher-order units that are themselves inherently stable and coherent. This then allows these new units to be further integrated with this pattern of integration and emergence continuing.

Openness in systems leads to transaction, relationships, and strategic decision making. This is consummated in union and forms the basis for hierarchy. It is no wonder that this pattern, which is the basis of vitality in all life forms, has been widely adopted in social advancement, as well as in technology development, as we observe in our next example.

Military forces, such as an army, exhibit hierarchy with the basic unit being the soldier and progressively higher forms of organization then being squad, section, platoon, company, battalion, regiment, brigade, division, corps, and then the army (of, say, soldiers) itself at the top. Each unit has a person in command, and these are expected to form leadership groups to bring the necessary command, control, and communication structure to the hierarchy. Chains of command, if strictly adhered to, can be lengthy, time-consuming, and burdensome. Breaking the chain of command has consequences, which, as we saw in the previous chapter, can be countenanced under the influence of pragmatism.

## I'M LOOKING THROUGH YOU

Openness and hierarchy work well today, but only if the unit formed from the integration of components, be these cells, soldiers, or circuits, is itself stable and coherent. This enables the unit so formed to become part of a continuing integration process up the chain, as hierarchy and openness do their work. This is where emergence comes in.

How do you know you have a system in the first place? Or, for that matter, a part? What do these objects mean to you if they are not definable, discernible objects with an existence entirely of their own? It's true that a system has composition, and the same is indeed true of a part. Looking into something is what we do to find its structure and its process. But its function attaches to the system itself, as a system. A platoon belongs to a company and that matters to the platoon. But the platoon matters to itself, and that is demonstrated by the soldiers who make up the platoon having a devotion, service, and loyalty to the platoon that transcends their own self-interest. The platoon can be good for the company only if it is first and foremost a platoon. And a soldier can be good for a platoon only if he or she is first and foremost a soldier. This duality is somehow subtended without doing injury to the component, as a constituent system, and in no way detracting from the system, for that component is prepared to lose its own identity. When all of this is in place, then emergence is said to be present, giving the system, be it platoon, soldier, or company, its right to existence unto itself and as part of a wider world.

In this book, we are urging you to consider the existence of worlds of systems that comprise systems we know about but that, in exercising their right to exist, become members of wider systems that somehow have escaped our attention and appear to be the province of no one. These worlds have emerged. They exist; they are very real; and they are highly impactful. The fact that we have somehow turned a blind eye to their existence makes them all the more potent and their unanticipated collapse all the more impactful, particularly on the systems we do know about. Navigating these worlds first takes recognition of their existence, and thereafter a skill set that deploys our Conceptagon.

# SYSTEMIC MAPS: SYSTEMIGRAMS

# CHAPTER 15

# WHAT . . . ?

## WHAT IS A SYSTEMIGRAM?

The first thing to say about a systemigram is that it is a diagram. While there are all kinds of diagrams based on content, function, and form, we can use as a good example the map of the London Underground, or the Tube, as it is often known. The London Underground system consists of many different entities, but the map serves mainly to indicate two of these: the stations, where passengers enter and exit the system, and the train lines on which trains operate carrying passengers from, say, Highgate to Westminster. The stations are mainly indicated by circles, and the sections of line on which the trains run, by straight lines connecting the circles. The different train lines such as Bakerloo, Central, and Northern are indicated by different colors. Stations are identified on the map by adding their names to the circles. There is an exception. Some stations are identified by a short line at right angles to the line

*Systemic Thinking: Building Maps for Worlds of Systems*, First Edition.
John Boardman and Brian Sauser.
© 2013 John Wiley & Sons, Inc., Published 2013 by John Wiley & Sons, Inc.

indicating the train line. These are stations that are unique to that train line, for example, Highgate. Stations where you can exit one train line and get on another, maybe one of several, are indicated by a circle. Westminster station, for example, is common to three lines: Circle, District, and Jubilee.

The map itself is a simple yet informative piece of design. It depicts a complex system yet in a relatively simple manner. It contains a host of information yet it is elegant. Regular users of the system hardly bother to consult the map, but when they go wrong, for lack of concentration or because their regular journey is unexpectedly interrupted, then the map is important. Tourists find the map essential. Considering how many people from all around the world use the Tube, the map is a wonderful piece of information brilliantly composed as a diagram. It has flaws in the sense that one might feel the need to use the system to get from one station to another indicated by the geography of the map, when in fact it would be quicker to walk since the two stations in question are more or less around the corner from each other. The regular knows this; the tourist may be misled. In spite of this minor failing, the map is a marvel. As a diagram, a map faithfully represents the system; in our example, the purpose is to navigate central London and the far reaches of that capital city. The systemigram as a diagram must also be a faithful representation. But of what? We shall see!

There is a second thing to say about the systemigram and that is that it is a system. So a systemigram is both a diagram *and* a system. Now the London Underground is a system, whereas the map of the London Underground is a diagram. How can something be *both* a diagram *and* system? Well, there are all kinds of systems. The London Underground is certainly a system. Attempting to say what this system is can be more difficult than might first seem. One can say that it is "a mass rapid transit system serving millions of people on a daily basis by providing a safe, convenient, and inexpensive means of getting to and from work, navigating the sights of London, and transferring between two London airports and the major city railway stations." That is just one example, which would be called a system description. Others are possible and valid, making use of more (or fewer) words, depending on the point of view. The trains that carry passengers are systems in their own right. The company that makes these trains is a system; some would call that kind of system an *enterprise*. The various pieces of equipment that provide the signaling mechanisms, indicating to train drivers whether it is safe to proceed or that they must stop, is a system. That system is also a part—part of the London Underground system. The people who work for London Underground constitute a system that also forms part of the London Underground system. So there are many kinds of systems. Some are physical objects, some are groups of people, some comprise solely software, and others are even more abstract such as a body of knowledge, for example, algebra. *If systems are so catholic, why should a diagram not be a system also?* To answer that question, we need to think abstractly for a little while as to *what makes a system a system*, whatever it actually is.

## WHAT IS A SYSTEM?

The simplest definition of *a system is a collection of parts and relationships that forms a whole that is somehow different, having its own personality as it were.* A familiar way of saying this is, "*A whole is more than the sum of its parts.*" There are all sorts of explanations for this difference and for this distinct wholeness. One explanation is that *the parts behave differently when they are actually parts*, that is, belonging to the whole, from when they are outside the whole merely designated to become parts. Another is that the relationships between parts are dynamic or continually changing and that this dynamism affects the parts when they themselves are belonging to the whole. A circle is a circle. But it's more than a circle when it's part of the London Underground map. It becomes a place, an origin, a destination, an exchange point. It is, in the context of the map, both a subway station and a feature of London. So also are all the parts and relationships in that map, which as a whole is a beautiful piece of industrial design and a work of art. Maybe the map of the London Underground is a diagram *and* a system? Does that make it a systemigram? No! Okay, then. So what makes a systemigram what it is? Something that is both a diagram and a system *and* unique in terms of its own design and artistry.

Let us return to the system description of the London Underground system. You don't have to look back; here it is:

> A mass rapid transit system serving millions of people on a daily basis by providing a safe, convenient, and inexpensive means of getting to and from work, navigating the sights of London, and transferring between two London airports and the major city railway stations.

Here is an interesting question: "Is this description a system?" Can a relatively straightforward sentence be considered a system? Let us examine that claim. First, it is a sentence that is a collection of words that make sense. So the words are parts. Where are the relationships? Well, *the order of words is important, and so ordering, governed by the rules of grammar, sort of implicitly denotes relationships*. Also, for the sentence to make sense, not only must the ordering be governed by rules, but the meaning that is formed by the combination of words must make sense. So the concatenation of words "The cat mat sat on" is not a sentence; it does not make sense. It cannot be a system. But "The mat sat on the cat," while being a sentence, really does not make a system since it is hard to understand the sense that this sentence makes. The familiar sentence "The cat sat on the mat" is grammatically sound and makes perfect if uninteresting sense. We might regard it as a system. But there are more interesting ones, and the system description earlier points the way.

Making sense is important and has many more implications for understanding what makes a system than might first be realized. We often ask our graduate students this question: "Who came after Harry Truman?" A few, mostly the younger ones, don't know the answer and may even struggle to know the gentleman was the 33rd President of the United States. All the rest quickly converge on the answer of Dwight Eisenhower. When we tell them they are wrong, we have to act quickly to dispel the confusion. We tell them that our answer is Doris Day! One or two get it, the rest are told to Google "Harry Truman Doris Day" and quickly discover the hit "We Didn't Start the Fire" by Billy Joel.

Our students then realize that we make sense of the words "came after" via the context of the lyrics of that hit song, and not the succession of U.S. Presidents. Sense depends on context; answers to questions depend on context. Pictorially, *if the question is depicted as a circle and the context is the exterior of the circle, then the answer to the question is the interior of the circle. Change the outside and the inside must change in response.* That's classic adaptation. Same question, new meaning, different answer.

In thinking about the meaning of a system, what this illustration does is to make us focus on important concepts that naturally go together like *boundary, interior,* and *exterior.* We can use these concepts to further ensure that a description of a system is itself a system, albeit an abstract system. Now all we need to do is find a way to turn the system description, the sentences, into a diagram. Then it would be both a diagram and a system. It would become a systemigram. But *to make a worthy systemigram, one must first assemble an equally worthy system description.* There are other system concepts that can help us to expand and develop our thinking about a system description. These other concepts can help us to turn a simple statement into one that is more fully developed, has greater richness, and tells us more about the systems it seeks to describe. This development process enables us to better understand that system (of interest) with a view to improving it or using it more successfully, depending on the interest one has in that system.

## WHAT IS A SYSTEM DESCRIPTION?

A very simple system description is *The cat sat on the mat.* This sentence can be transformed into a diagram (Systemigram 15.1). It's not the most compelling picture that has ever been created, but it's not entirely without interest, an interest that grows as we understand the interplay between system descriptions, systems (that are being described), and systemigrams (that portray the systems being described).

There appears to be two parts and one relationship in this diagram, making it in principle an elementary system. The parts are

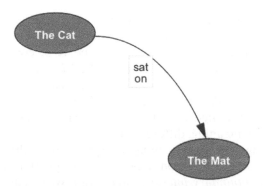

**Systemigram 15.1.** The Cat Sat on the Mat

**Figure 15.1.** Sophie Sitting Comfortably

depicted by nodes (ovals or ellipses) and the relationship by a link (a directed arc pointing from the top left node to the bottom right. The labels in the nodes are the nouns in the sentence and the label on the link is the verb. This mapping is the basis for creating a systemigram. The diagram, as a whole, could be depicted by a photograph (Figure 15.1).

Because it's possible to link this picture to the diagram, *the diagram has to mean more than being just a couple of nodes, a link, and a few words*—at least to some people.

But the structure of the diagram, though elementary, is interesting. It's a beginning. The process of building it is also interesting, and the function that the diagram points to, that of a lovable creature entertaining us by relaxing, is even more enthralling. The words *structure, process,* and *function* are another trio that belongs to the set of systems concepts that help us shape our thinking. This trio or triple can now be added to the two we have already introduced: *boundary, interior, and exterior* and *parts, relationships, and whole.*

If we wished to develop a richer system description of the London Underground, these three triples are very useful prompts to help us do that. For example, we might ask the following: What is the *structure* of the London Underground? What are the *processes* that these structures fulfill, both internally and collaboratively? What are the *functions* that the Tube serves, for example, making it safe, inexpensive, and convenient for people arriving at London City airport to transfer to Euston railway station? What are the *parts* of London Underground, and the *relationships* between these parts? What does the *whole* mean in terms of revenue, expenditure, maintenance, reliability, and so on? Finally, what is the *boundary* of this system? Is it merely the extremities of the stations, or is it the statutory responsibilities of those charged with its operation? Discovering what is on the inside of this boundary or the *interior* of the system is relatively simple, one would think, having defined the locus of the boundary, but what is on the outside? What is the *exterior* of the system? This is not just "everything else," but, in particular, it's those things that matter to the system, things that can affect the makeup of the interior just as Billy Joel influences Harry Truman's successor.

*A basic question to ask of any system description is what is the interest in the system being described.* Okay, what are the interests in the London Underground? One person, an engineer, is interested in the safety of what is essentially a system for moving millions of people per day tens of millions of passenger miles deep beneath the city of London. Another, a woman entering the autumn of her years, is concerned by the complexity of the map and the fear of getting on the wrong trains and therefore being prematurely eternally interred!

*There exist in reality multiple perspectives.* Companies are interested in contracts to keep stations clean, to supply vending machines for passenger tickets, to provide signaling equipment for train drivers, to offer training courses for personnel management, and so on. The system is extensive, multifaceted, and repercussive while simultaneously serving as a crucial infrastructure to the economy of one of the world's major centers for tourism and financial services. London, home to millions, needs the Tube for its transport needs, just as any engine needs oil. This complexity is a challenge to the authors of a simple description of that system. However, there is hope.

*Just as the map of the London Underground need not be as complex as the system it portrays, neither need a system description be as complex as the system it describes.* True, there must be a reflection of the immense variety of perspectives, issues, concerns, and needs, but so also must there be a parsimony (or meanness) applied to capturing this variety in a description. *A system description is a model of the system, and all models are wrong but some are useful. What guides a system description to be a useful model is the judicious application of system concepts*, and to that end a new triple, namely, *variety, parsimony, and harmony* helps greatly.

What are the ways in which this triple can be applied to the formulation of a description of the London Underground? Let's start with variety. We have already pointed to the number of different perspectives that are taken with regard to this system. For sure, there must be hundreds if not thousands of entities with a valid perspective into this system, not to mention the millions of perspectives of the traveling public. Variety, it would seem, is not a problem, but dealing with the overwhelming nature of this variety surely is. That is where parsimony enters. *Somehow or other this enormous variety must be reduced to a manageable size, but the goal of manageability must not bring the forces of variety and parsimony into conflict even though on the surface this seems inevitable.* That is where harmony comes in. Harmony is the expression of desire for equilibrium between two opposing forces that results in these forces working together to achieve a more desirable result than if they worked separately or not at all.

For our purposes harmony will bring the variety of perspectives on the London Underground system into a clear focus in which each interested party can not only recognize its own involvement and be able to see the contributions of others, but also, perhaps for the first time, discover how all these efforts, all rendered with good intent, impact each other and not always beneficially but often adversely, thereby leading to the kinds of problems and challenges that the system as a whole presents.

It is as though *the system description becomes a screenplay with roles, characters, dialog, intent, reaction, and all those other features of a narrative being interwoven.* The author of the system description is not merely describing a static system but rather capturing the action, the system in motion, with participants involved, knowingly or unwittingly, in a plot that often confuses, sometimes entertains, and always thrills. The ultimate goal is to engineer meaning for the viewer (as well as the participants) and hopefully a happy ending.

## WHAT IS A SYSTEM OF INTEREST?

The fact is that there is no single story of the London Underground; there is a multitude of them. There are stories about its history, covering its origins, ownerships, and operations. There are stories about specific developments, such as the Jubilee line extension or the new station at Heathrow's Terminal 5. There are stories about disasters, such as the Moorgate crash in 1975 or the King's Cross fire of 1987. *All of these stories bear witness to specific systems of interest, and each specific system of interest is deserving of its own system description and subsequently of its own systemigram.*

The London Underground system has given birth to many stories and to many systems, and while it is possible to describe it in words, however many or few, that system description is neither greater nor less than any one of a host of other system descriptions, each of which will have connections with the London Underground. The question being posed therefore is, *"What is your interest?" Once that is settled, then we have effectively begun to define a system boundary*. Development of a relevant system description can begin, and this will in effect be the interior relative to that boundary. Yet now we know, courtesy of our system concepts, that *beyond our interest, there lies an exterior, and as much as we might choose to ignore this, having settled our interest, it must somehow become part of our system of interest*.

We are now going to move above ground but remain in the United Kingdom and settle on a specific system of interest, a slightly different type of rail system that became subject to privatization or deregulation. The London Underground system has experienced several kinds of ownership itself, and one of these included an element of privatization, when in January 2003, it began operating as a Public–Private Partnership (PPP), whereby the infrastructure and rolling stock were maintained by two private companies (Metronet and Tube Lines) under 30-year contracts, while London Underground Limited remained publicly owned and operated by Transport for London (TfL), which replaced London Regional Transport (LRT) in 2000, a development that coincided with the creation of a directly elected Mayor of London and the London Assembly.

As far as London Underground is concerned, PPP was controversial from the start. Supporters of the change claimed that the private sector would eliminate the inefficiencies of public-sector enterprises and take on the risks associated with running the network, while opponents said that the need to make profits would reduce the investment and public-service aspects of the Underground. The scheme was put in jeopardy when Metronet, which was responsible for two-thirds of the network, went into administration in July 2007 after costs for its projects spiraled out of control. The case for PPP was further weakened a year later when it emerged that Metronet's demise had cost the U.K. government £2 billion (about $4 billion). The five private companies that made up the Metronet alliance had to pay £70 million each toward paying off the debts acquired by the consortium. But with the U.K. government's agreement in 2003, the companies were protected from any further liability. The U.K. taxpayer therefore had to pick up the rest of the tab, which further undermined the argument that the PPP would place the risks involved in running the network into the hands of the private sector.

Getting the economic architecture of PPP correct had somehow been overlooked, and it is this very quandary that bedeviled the privatization of British Rail. Our system of interest is that very subject matter, and in particular, it is most sharply focused by an article that appeared in November 1998 in the Sunday *Times*, written by Gerald Corbett, at that time CEO of Railtrack. His protestations were sensibly combined with a way forward, and our analysis of his article enabled us to produce a system description, for which we prepared a systemigram, shown as Systemigram 15.2.

In this opening chapter, we intended to tell you what a systemigram is. Now we want to tell you what *this* systemigram is because it is certainly more complicated than *The cat sat on the mat*, and we want you to understand not just this one, but others that can get even more elaborate.

## WHAT IS *THIS* SYSTEMIGRAM?

We have said that our system of interest is the economic architecture of the privatized rail industry in the United Kingdom around the time of the late 1990s and how this is failing, why it is failing, and what remedy is being suggested (by the leader of one of the stakeholders, Railtrack), which, given the collaboration of other stakeholders including the U.K. government, might put the industry back on track (so to speak).

That statement of the specific system of interest defines a system boundary, and you would fully expect to find in the interior such explanations as the structure of the industry, the flow of money around the industry, the roles and responsibilities of stakeholders, the reasons for a failing architecture, and the suggested way forward. What matters to all of these things is the context or the exterior of this system boundary, and that is where we now start in explaining what this systemigram is.

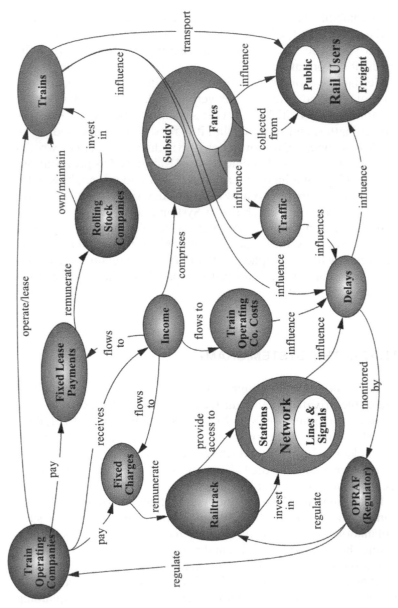

**Systemigram 15.2.** U.K. Rail System

Deregulation of the U.K. rail system had three key objectives:

- to cut the railways' government subsidy
- to boost traffic—in 1995, there was zero growth despite road congestion
- to improve punctuality—a better service encourages road users to switch to rail.

The transformation of the rail industry was intended to achieve these objectives by setting out a corporate landscape and overlaying upon this an economic architecture. The pieces of this landscape are made up of four kinds of enterprise (a particular example of a system, you will recall). First, there are the 25 companies that operate the services for passengers and freight, known as the train-operating companies (TOCs). Each of these receives a steadily declining government subsidy, and fare income that grows as traffic builds up. A second kind of enterprise is known as a rolling-stock company (ROSCo). Each ROSCo owns and overhauls its trains. A third element in the picture is Railtrack, which owns and maintains the tracks, signals, and railway stations. Railtrack is in effect a supplier, of network capacity, to the TOCs. Finally, there is the government-appointed regulator of the system, the Office of Passenger Rail Franchise (known as Opraf). Among its duties are the granting of licenses to the commercial elements in the rail system and the application of rewards and/ or penalties for performance of these commercial service providers (Systemigram 15.3).

The economic architecture can best be described in terms of the money flow across this landscape, as determined by the transformation process. The TOCs pay largely fixed charges to use Railtrack's lines and make fixed lease payments to the ROSCos. The fixed charge to use the lines was set to enable Railtrack to maintain and renew its network, to upgrade all 2500 stations, and to eliminate the big investment backlog. At the time of privatization, Railtrack's lines and signals were responsible for 65% of delays. Yet its income was largely fixed, giving it no incentive to improve punctuality. In response, the government gave it a performance regime with

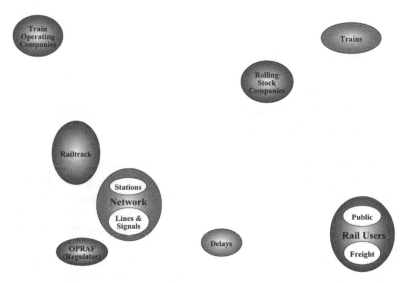

**Systemigram 15.3.** U.K. Rail System—"The Corporate Landscape"

strong incentives to improve its network. The view was that the TOCs would not require incentives, because if they performed badly, their fare income would fall as passengers migrated back to their automobiles.

Opraf introduced performance regimes, but Mr. Corbett argued that the incentives were weaker than for Railtrack. In 1997, Opraf paid £13 million net to the TOCs for performance, an average of £500,000 (about $1 million) per TOC, but all of this was then paid by the TOCs to Railtrack under its performance regime. The economics of a typical TOC are something like the following: ticket sales produce 60% of revenue and the subsidy 40%. Of the outflow, 40% goes to Railtrack, 18% to the ROSCos, 20% to the staff, and 18% to cover other costs. This leaves a 4% operating margin. In this situation, what do you suppose the managers of a TOC will do? Which levers will they pull to set profits in the right direction?

TOCs will do everything they can to run more trains and attract more passengers, since almost all the extra income will pass through to the bottom line because most of the Railtrack leasing and labor costs are fixed (Systemigram 15.4). Second, they will cut their costs because the state subsidy is falling by about 15% a year. This means sales must grow and costs must fall for the TOCs to stand still. Meanwhile, the theory was that managers would still aim to run the trains on time because "if they did not, they would lose passengers." Things did not turn out that way, however.

Mr. Corbett writes:

The industry has grown faster than the architects of privatization imagined three years ago. Passenger miles have risen 16%, revenues by more and there are about 10% more trains on the network. The TOCs have cut costs—a 5% cut in jobs is typical. Their share prices have risen as higher sales and lower costs have come through to the bottom line. Railtrack has also performed, with delays caused by it and its contractors more than 40% down over three years. Infrastructure is now responsible for only about 45% of delays. There is more to be done, particularly on the Great Western lines, but progress has been made. Railtrack has more than doubled investment—to £1.25 billion last year and £1.45 billion this year. By the new year (1999), the British Rail backlog will have been almost eliminated, and the station upgrade program will be half finished. By 2001 Railtrack will have spent £1 billion more than the regulator assumed when setting its access charges in 1994. The state subsidy fell £285 m this year to £1.6 billion and will be £926 m in 2003–2004. Privatization has delivered what its architects intended. But with success has come a problem—poor punctuality. Railtrack, responding to its incentive regime, has cut delays caused by tracks and signals but the TOCs, with some exceptions, have not made similar progress. This is because their economic regime is potentially lethal for punctuality. Trains are added, jobs are cut, punctuality pressures mount, but the growth hides any loss in fares due to the poor performance. But the problem will not go away. As the subsidies fall, the pressure will build and a recession would intensify it. Some TOCs will be unable to keep investors and customers happy. The economic architecture will continue to drive them to actions that cause delays.

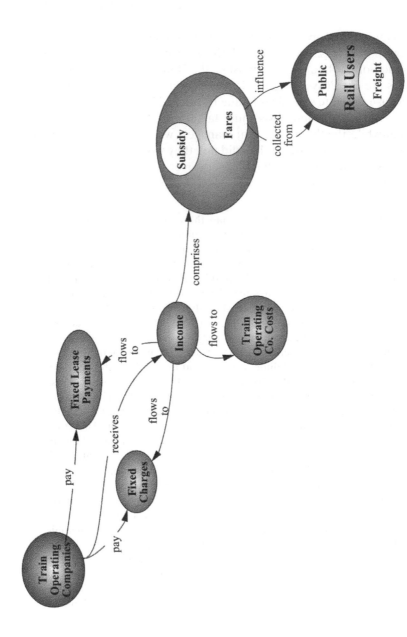

**Systemigram 15.4.** U.K. Rail System—"Follow the Money"

What can be done? Mr. Corbett does not recommend waiting 5 years when the TOCs' new franchise agreements start. Passengers have suffered enough delays. He proposes a new regime to include:

- Bigger incentives to make punctuality a real profit lever for the TOCs.
- An access charge related to sales to give Railtrack an incentive to encourage growth.
- Longer franchises to encourage TOCs to order new rolling stock.
- Government assurances that the industry will come through the regulatory review strong enough to fund investment.
- Existing subsidies to be realigned to provide incentives for desired outputs. (Systemigram 15.5)

In closing, Mr. Corbett writes:

The privatized industry has much of which to be proud: good growth in passenger and freight traffic, billions of pounds of private capital in the sector, entrepreneurial managers with new ideas, steadily improving safety, a big rise in infrastructure and rolling-stock investment and a rescue for the Channel tunnel fast link. But the question remains: must travelers wait until 2003 for punctuality to be addressed? They deserve better.

What is to say that Mr. Corbett's conclusions are correct? That he does indeed offer a remedy for treating the ailments of the system as a whole and not merely the symptoms of its parts? After all, he does and must have vested interests in the enterprise he leads, Railtrack. But this would be true of any of the leaders of the other enterprises that make up the rail transport enterprise landscape. On that basis, none would be heard even though anyone might dare to speak for all. This is not a trivial problem. And no matter how seductive a simple answer might look, for example, renationalization of Rail UK, it would be unwise to be satisfied by a simplicity eagerly grasped in the face of such complexity.

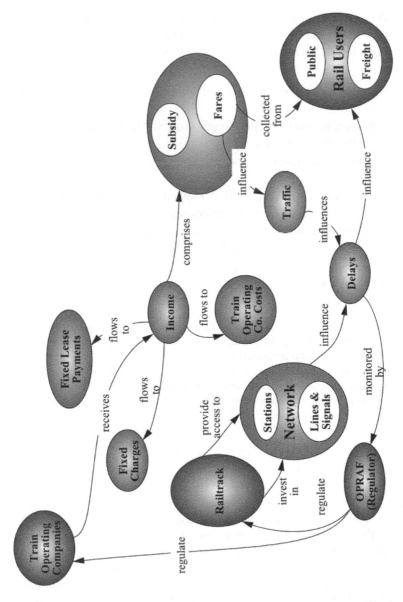

**Systemigram 15.5.** U.K. Rail System—"Way to Go"

However, now that we are in possession of *this* systemigram plus a basic knowledge of what one is, perhaps *a medium is opening up for us that helps orchestrate the debate among problem owners* whereby the narrative they are following, sometimes by improvisation, given to them by others, becomes clearer and a happy ending is arrived at.

# CHAPTER 16

# WHY . . . ?

## WHY DO SYSTEMIGRAMS EXIST?

Some 20 years or so ago, the emphasis in the European Economic Community (EEC) was on integration. To attempt political integration of sovereign nations with the goal of forming a federation to rival that of the United States was, at that time, manifestly infeasible. However, to achieve improved collaboration between nations in areas such as Research and Technology Development (RTD) appeared practical and realistic. This would, or so it was thought, minimally bring together the undisputed talents of engineers, scientists, and technologists to create a critical mass that would rival, say, Silicon Valley. It might also prove a useful test bed for the future political integration that some dared only dream and speak of furtively.

Accordingly, the EEC developed a Framework Program to sponsor the best teams and ideas in ways beyond the reach of

*Systemic Thinking: Building Maps for Worlds of Systems*, First Edition.
John Boardman and Brian Sauser.
© 2013 John Wiley & Sons, Inc., Published 2013 by John Wiley & Sons, Inc.

national initiatives. Unsurprisingly, the ambitions of faculty across Europe, combined with the avarice and cunning of corporate executives to secure matching funding of their internal RTD expenditures with EEC monies, ensured a plentiful supply of project ideas to compete for funding.

One such project rejoiced in the splendid acronym of ATMOSPHERE— Advanced Tools and Methods for System Production in Heterogeneous Extensible Robust Environments. This is easy for us to remember because we were engaged by the EEC to help the project achieve its goals, an unforgettable experience lasting 4 years but with consequences for us that have endured to this day and look likely to continue in the foreseeable future. It was in that atmosphere that systemigrams first saw the light of day.

Our first task was to find out what the project was all about and why in particular the EEC had funded it against others and to a much greater amount than any other project. The scope, scale, and challenge of the ATMOSPHERE project drew together the great and the good across Europe—Bull, GEC Marconi, Olivetti, Philips, and Siemens were the big players, with an extended cast of lesser characters drawn from the small and medium-sized enterprises (SME) community, much beloved of the EEC and accordingly appropriated a kind of affirmative action status.

The document that had formed the basis for the EEC's evaluations and subsequent funding allocation became our reading matter of special interest for several weeks and allowed us to form our own synopsis, which comprised, in our judgment, the significant features with which the project would be intimately concerned. The major players wanted their relevant engineers to develop systems engineering tools that would aid in the design of specific product lines, for example, industrial control systems. The wide variety of tools (and methods) that the players required posed a challenge when it came to the matter of developing a single unified environment that all would use to create these diverse tools. Would it not be better for each tool developer to go his or her own way? That, according to the EEC, failed the integration test. To get the money they needed, companies subjected their specific tool developers to sign up to a single integrated environment, one that supported not only heterogeneity (of tools and methods) but also extensibility (future unforeseeable tools) and robustness (i.e., the environment would not break under the stresses of diverse unforeseeable tool development needs). Quite a challenge! Today's equivalent would be Apple's environment to support development of applications for the iPhone or iPad. Much has happened over the past 20 years!

Our initial encounters with the principals of the project workforce left us in a quandary. It did not appear that anyone understood the architecture of what was required. It was as if they had not read the document that had assumed biblical proportions in our lives for the previous several weeks. If they had, their personal interpretations differed widely from our own, and we wondered if we had got it all wrong. And why should we not have? They were the experts; we were merely troubleshooters hoping to give sound advice to folk who had invested huge sums of money for a good cause. Naturally, we went over those 100-plus pages again, but we knew that to give a good account of our understanding of ATMO-SPHERE, we had to find a way of exemplifying this in a noncontentious (user-friendly), simplified (but not simplistic), and hopefully extraordinarily remarkable fashion. Fine, but how?

Our previous researches had taken us on an interesting journey through the artificial intelligence (AI) field courtesy of Intelligent Knowledge-Based Systems (IKBS, aka Expert Systems). This experience had opened our minds to semantic networks, a medium for representing knowledge in a formal way that could then be used as a basis for both computation and cognition. We had extended this medium a little, bent more toward cognition of complex systems in simplified terms. It seemed to us that ATMOSPHERE represented an excellent test of our ideas, make or break conceivably. The question we asked ourselves was this: "Could the significant features of the ATMOSPHERE project, as we had come to frame them, be portrayed in a simple yet integrated fashion that would point to possible pitfalls and provide a lodestar for task navigation by the team and a monitoring device for the EEC?"

Thus we created our very first systemigram for public consumption. Had we known at the time what a landmark achievement this would become, we would have taken better archiving care. Sadly, this was not the case, and we are unable to exhibit it today. This perhaps only adds to the intrigue, for that diagram, which we believed also to be a system, represents breakthrough thinking of the first order. Here now is the tale of what happened.

We were invited to a formal review of the project to evaluate the progress made, or lack of it, and to consider whether funding should continue, increase, or cease. To all intents and purposes, it was make or break. The appointed EEC agent, who had especially embraced the ATMOSPHERE concept and who knew its conceiver very well, the author of the document we had read and who had moved on to other ventures, chaired the meeting. He showed the assembled gathering two documents, each of a single page. One was a fax sent by the project team summarizing progress. Its brevity did not reflect a succinctness of achievement to be envied, but rather the paucity of progress and a hopeful statement that "things would improve." Dramatically, one could say that it was a $6 million straw man. The other was our systemigram. This had been procured for significantly less but was now being vaunted to have much value.

As he waved these two white sheets of paper, holding one in each hand, seated at the head of the long table around which his paid servants had gathered, every eye was fixed on him, and you could hear a pin drop. What was coming? He castigated the project team in a manner we will never forget. It was dignified abuse of the first order. He coupled his unerring rebukes with unbridled praise for our work. We wondered, "If only this guy's costs for each sheet had been switched, as his remarks seemed to indicate, that would be more just." These two plain pieces of paper began to symbolize white flags of surrender. But from whom and to whom? Was the team to give up in abject failure? Was the EEC to abandon its main project, the jewel in the crown, an enterprise "too big to fail"? Or was there a glimpse of hope? Could the team recognize in our systemigram a mission that had once been fully embraced by the EEC and was still worth much to them if it could succeed? Could the team debate the fallacies of that mission and find a better way, one that still exemplified the collaboration that the Framework Program espoused and deliver the prototypes that would become European products to compete on a global scale? *Our systemigram had spawned these strategic questions and in so doing initiated our journey into systemic media for problem owners.*

We learned that whatever the outcome, whether the project continued, was abandoned, or changed, our systemigram had succeeded. Of course, much work had gone into understanding ATMO-SPHERE or least as far as the proposal document was concerned. But to take that understanding and portray it in a comprehensible form, one that would guide future decision making, was a breakthrough. We needed to build on that. And so we got the chance. The project continued for two more years. Annual reviews featured systemigrams prominently to illustrate convergence and advise on progress. The project director, when he finally left the company that had appointed him to the task, set up his own consulting operation that advised other EEC-funded projects on how to use systemigrams! *Why do systemigrams exist?* Because a jewel in the crown had slipped and it needed to be restored. Why do they continue to exist? *Because there are some problems that, regardless of all the solutions being proffered, are not that simple. But a simple explanation of that complexity is highly desirable.*

## WHY ARE SOME PROBLEMS COMPLEX?

We learned some important lessons from working on the ATMO-SPHERE project and we believe that these contribute to our understanding of the culture that permeates the mystique of complex problems. *The first lesson is that people get into trouble or difficulty when they are actually striving only to do their best.* It's not that their best is not good enough, though that may be true; it is rather that for some reason or other the good that they do has the counterproductive effect of harm.

Mathematically, subtract 8 from 10, and the result is 2. What could be simpler? But if the 10 are Al Qaeda operatives and the subtraction translates to death, how many are left? Is it zero because the survivors have a change of heart? Or is it 20 because the friends and relatives of the slain are emboldened to join the cause and swell the ranks of jihadists? This is the math that once confronted General Stanley McChrystal. This is no academic nicety. It is *a problem of considerable complexity and yet a simple solution is the continuing expectation of us all.*

Cats as you know are possessed of the need to catch and kill small birds. We have a cat, her name is Delilah, and she is no exception to her species. She has observed that birds gather aplenty on our bird feeder in the backyard. Though it rests atop a 7-ft wooden pole, this obstacle is no serious hurdle to the ever-resourceful Delilah. She scales the heights and sits patiently for the birds to come. Their continuing absence confuses her, and yet she waits. Finally, she returns to other more reliable sources of food. The birds return to feed as she blinks dubiously at them while enjoying a bounteous supply of "yellowfin tuna Tuscany in a savory sauce with long grain rice and garden greens." What did she do wrong? She doesn't know. We know that because from time to time she repeats her counterproductive terrorism of our feathered visitors.

*When people don't know what else to do, they do their best.* That was true of the workers in ATMOSPHERE. A host of computer geeks knew how to develop tools. The goal was to develop a common environment in which tool development could proceed more productively and which would permit easier access to a broader community of tool developers. When this rallying cry went unheard, or possibly unenunciated, the tool developers did what they knew to do. As far as the EEC was concerned, they could do that by all means, just not on their dime! Tool development effort went the way of Delilah's search for birds.

*A second lesson is that many people don't ask questions, preferring to get on with what they are told, or get on with what they know when they are not told (or don't hear).* Asking questions can be dangerous. You risk being labeled ignorant or insubordinate. This mindset breeds a culture of fear, and those who deliberately cultivate such a culture exercise a control that is both perverse and ultimately self-destructive. Such an outlook is not confined to lower echelons; it is practiced all the way to the top. People out to close a deal with a customer play it safe; they do what the customer wants even though they might know it can't be done or will cost their company too much money and grief. It is hard to turn down money. People always make something work or they move on to other deals. Either way, money talks.

These are the kinds of influences that contribute to complexity. It was certainly at the root of the ATMOSPHERE project. In the end, the original mission that had been envisioned had to be changed. But the change was arrived at by rational discourse, as well as due recognition of the prevailing pragmatics. That discourse was well served by our systemigram. *That was a third key lesson. By showing all involved, customers, contractors, suppliers, and others, the big picture in a simple yet integrated fashion worked. It brought out the sense of the original mission and gave the assembled team, who had not been party to setting that mission, every opportunity to comment on it using their acclaimed expertise.* Arriving at a consensus of the infeasibility of that mission was achieved rationally rather than emotionally, though emotion played a part. Stakeholders felt obliged to put forward penetrating questions and were well served by asking them.

*Complex problems didn't necessarily start out that way.* At one time, many were fairly simple. Over time, they became more complex, often as a result of ill-suited solutions. These measures did not eliminate the problem but rather compounded it. Problems are complex because the parts of them are massively interwoven, and often in ways that we do not observe. It is the web of relationships between parts that makes a problem complex. The eight slain Al Qaeda members are not only connected to their cell; if that were, then 10 minus 8 would surely equal 2. The eight are connected to family, friends, a cause, and a deep-rooted belief system. What effect does the measure of killing these eight have on these connections? Is it, in effect, the same as Delilah's measure to slay birds? *The answer might be that we just don't know. But in this case, what we know to do is to step back, to stand back and see the big picture.* That's what happened in ATMOSPHERE. Tool development ceased and the environment for tool development became the focus. Then the various processes for tool development could be calibrated against that focus and more informed decision-making made.

*It is risky to stand back. It appears to the uninformed observer that nothing is being done, while resources are still being consumed. But good is being served*: nugatory effort is suspended and the expenditure of wasteful resources downstream averted. *When faced with a complex problem, it must first be respected for its complexity. The problem owners need to be advised of its complexity.*

Previous solutions must be inspected for their contribution to the diminution or exacerbation of the problem. All of this requires a mindset that is contemplative and rational, which does not yield to "the fierce urgency of now," but which nonetheless makes up for time "lost" in contemplation by the avoidance of misspent effort downstream. Such a mindset deserves assistance and for that reason systemigrams usefully exist.

*A systemigram helps orchestrate debate among stakeholders and organize collective executive action pursuant to that debate.* A systemigram can help elicit new lines of interrogation, evoking questions that might otherwise go unasked, spurring lines of inquiry that can shed light on hidden complexity. *A systemigram portrays a big picture that normally eludes overburdened operatives and narrowly focused zealots eager to exhibit and practice their expertise. Once a big picture is portrayed, it becomes obvious, almost commonsensical. But until it exists explicitly, it remains a mystery, its absence sustaining a darkness overshadowing the search for answers.* That transformation, from the imperceptible to the self-evident, is part of the magic of systemigrams. One such transformation was furnished for a company in the GEC Marconi group, and this is the subject matter of our next section.

## WHY DOES *THIS* SYSTEMIGRAM EXIST?

In the 1990s, GEC was the premier engineering industry in the United Kingdom, with global reach in all product and service lines. It might have been likened to a three-legged stool with its three main divisions being defense and aerospace (GEC Marconi), telecommunications (GEC Plessey Telecommunications, or GPT), and power generation and transport (GEC Alsthom). GEC Marconi's only rival at the time was British Aerospace, and on more than one occasion, the opportunity arose for the former to acquire the latter. In the end, the reverse happened, forming today's BAE Systems. GEC retained the Marconi brand and subsequently focused uniquely on the growing global telecommunications market driven by the burgeoning Internet. When this company collapsed, the stool vanished from the world's corporate furniture. However, the work that was done in GEC Marconi involving systemigrams proved seminal and lives on today in BAE Systems, whose current CEO, Ian King, was thoroughly schooled in Marconi.

GEC Marconi was a collection, an alleged federation, of several companies, each one specializing in a particular product or service domain. At one point, that diversity was described by its Technical Director, Dr. Bill Bardo, as ranging from the seabed to outer space and from ships to chips. This languid attempt at poetry could not however disguise the reality that companies within the GEC Marconi group were fiercely independent, though not autonomous, since financial controls were unignorable and strictly adhered to, and often competed with one another for prime contractor status. A Managing Director would take great delight in winning a contract and offering his peers the chance to be a subcontractor on the contract they had bid for and lost to their now-gloating rival. Not unnaturally, this spirit of competition and rivalry bred a culture less than conducive to collaboration, and this in turn made it less likely that people within the different disciplines, for example, marketing, engineering, and finance, would happily cooperate even though they were on the same team. This state of affairs is not without its consequences.

We became involved with Bill Bardo (of GEC Marconi) as a result of a growing interest in our deployment of systemigrams and in particular because of a fierce urgency to reform corporate culture in order to make GEC Marconi wholly more effective. A specific system of interest became that of bidding for contracts. It may not seem a big deal; after all, isn't it simply a matter of receiving an invitation from a prospective customer that is then scrutinized and a response made in terms of a proposal as to what it's going to cost, what will be delivered, and what is the work plan that explains the how? That is being simplistic. The reality is that it costs money to formulate a bid. Of course, if the company never bids, it will have no work and soon go out of business. But by the same token, making an endless series of bids, few of which are successful, is another road to premature expiration.

The seniors in GEC Marconi were concerned that too many bids were failing and the costs of this failure were posing a seriously adverse threat to the bottom line. It was not as if the company lacked expertise in producing the deliverables required by the customer, but it did appear that there was a lack of competence in formulating the bids themselves. It was felt that some invitations to tender from customers could be safely ignored, while others given special attention. If there were to be a defined budget for formulating bids, how should this be allocated? Ideas turn on a grain of sand. Maybe some money should be set aside, a small proportion, not for formulating bids, but for formulating *whether to bid*. This grain germinated and eventually gave rise to a comprehensive end-to-end phase review model for the implementation of strategy across the whole of GEC Marconi.

*Our involvement was to portray the complex operations of the bidding process in a simple manner using systemigrams as our medium for storytelling, procedural verification, consensus building, and competency modeling.* The big picture for the bid/no bid decision process is shown in Systemigram 16.1.

This picture is really not all that complicated, but it nevertheless needs to be presented a bite at a time, so as to avoid unnecessary indigestion. The purpose of the bid/no bid activity is revealed in Systemigram 16.2. This states that the customer generates an Invitation to Tender (ITT), which is then dispatched (to a list of candidate contractors). Having been received by one such company, in the GEC Marconi group for our purposes, it is subjected to review and analysis by a Bid & Proposals Manager. It is his or her responsibility to develop a strategy for the bid, should it go ahead. That strategy is described in a bid plan, the principal features of which are broken out as a series of bubbles within the larger containing bubble (Systemigram 16.2).

This bid plan then becomes the responsibility of a second agency (who may in fact be the same person) known as the Project Bid Manager. His or her responsibilities are defined in greater detail in the next phase, should there be one, termed proposal production. One thing to notice here is the different nomenclature that aggregates around essentially the same artifact—tender, bid, and proposal. This is not confusing once it is recognized, but in many instances, an inconsistency in terminology can lead to ambiguity, confusion, and ultimately failure. To elaborate, the ITT is often known also as RFP (Request for Proposal) and RFQ (Request for Quotation). Words always matter but sometimes they matter little!

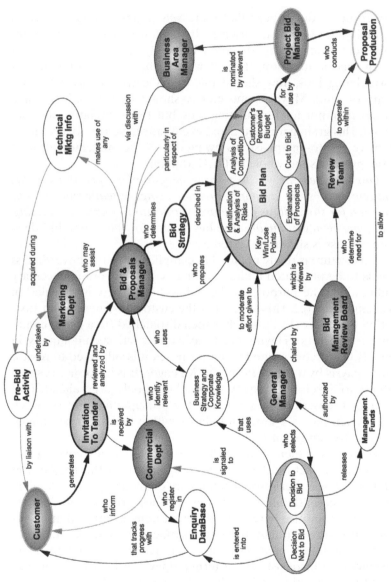

**Systemigram 16.1.** Bid/No Bid Proposal

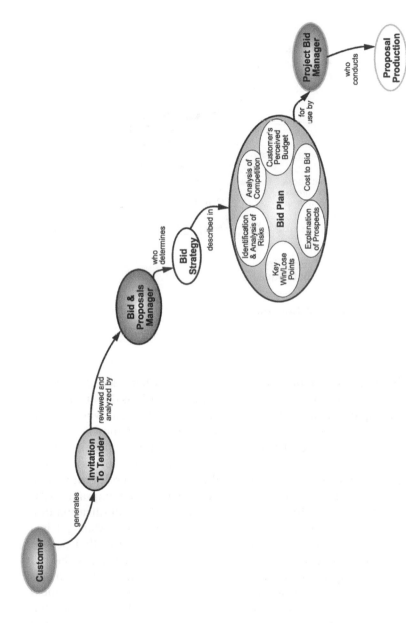

**Systemigram 16.2.** Bid/No Bid Proposal—"Main Thrust"

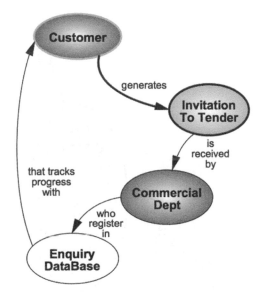

**Systemigram 16.3.** Bid/No Bid Proposal—"Initial Steps"

Now let us look in a little more detail at what needs to happen in order to come to a decision and how the various departments of the company wishing to make a decision congregate via the flow of relevant information. It is the duty of the commercial department to receive the ITT so that it can register it into an inquiry database that is used to track progress with the customer. That department also has the responsibility to identify a relevant person to play the part of Bid & Proposals Manager (Systemigram 16.3).

In Systemigram 16.4, we begin to see that the identification of a suitable person to play the part of Bid & Proposals Manager is a nontrivial task because the competencies required of that person are more fully revealed. This person, of course, prepares the bid plan, knowing what the relevance and importance of the various features are of the plan. In doing so, he or she must use business strategy and corporate knowledge to moderate the effort devoted to the plan's preparation. He or she would be wise also to make use of technical and marketing information that has been acquired during previous activities. Often, the ITT comes as no surprise. The customer is often a known quantity, the company having successfully done business with him or her before. It may even be that the ideas that underlie the ITT were planted by the company some considerable time before. It is the marketing department's role to conduct such customer liaison and to acquire this background information, which might proves useful for the bid plan. Systemigram 16.4 shows that the Bid & Proposals Manager must pay particular attention to the analysis of the competition and how much the customer is willing to spend.

We are beginning to see the importance of good working relationships between two departments normally concerned with distinct phases of the company's operations. Once the bid plan is prepared, it is reviewed by a bid management review board chaired by a general manager. It is he or she who makes the final decision—to bid or not to bid, that is the question. That decision likewise makes use of business strategy and corporate knowledge just as it was basic to moderating the effort to be placed into the bid plan. The decision is entered into the inquiry database and signaled to the commercial department, whose duty it is to inform the customer (Systemigram 16.5).

Let us now suppose that the decision is made to bid. The detailed activities that follow are now depicted in Systemigram 16.6. One thing that must happen is a release of funds to allow the next phase, proposal production, to proceed. This authorization is the responsibility of the general manager who chaired the bid management review board. That board's work is not ended. It must determine the need for a review team to operate within the proposal production phase, perhaps in similar vein to how the bid management review board itself operated in the current phase.

The last point of detail relates to the Project Bid Manager, who will be responsible for the proposal production phase. He or she is nominated by the business manager most relevant to the nature of the bid but who will wisely discuss that appointment in conversations with the Bid & Proposals Manager, the person who successfully prepared the bid plan from whose content downstream activities might benefit.

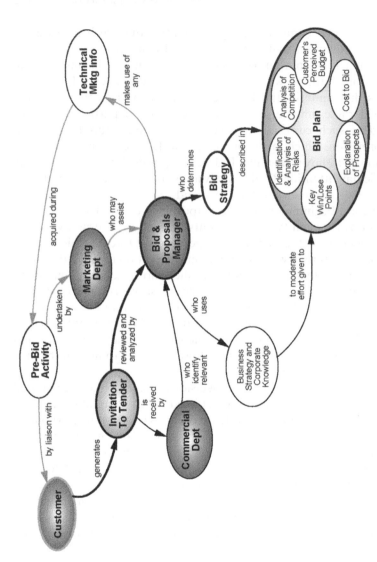

**Systemigram 16.4.** Bid/No Bid Proposal—"Background Information"

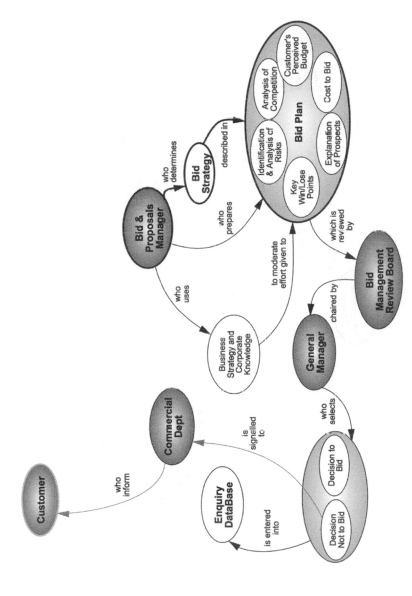

**Systemigram 16.5.** Bid/No Bid Proposal—"Decision-Making"

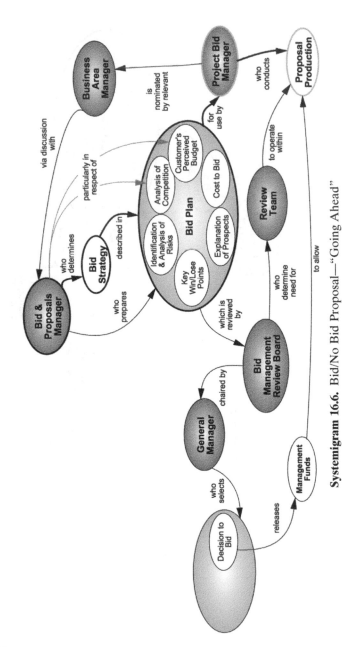

**Systemigram 16.6.** Bid/No Bid Proposal—"Going Ahead"

## WHY IS *THIS* SYSTEMIGRAM USEFUL?

Before we turn to the utility or otherwise of this systemigram, we should perhaps explain how it came about, given Bill Bardo's requirement to uncover the reasons for poor performance in the bidding element of GEC Marconi's overall strategy. Clearly, if you are engaged in an activity to discover reasons for failure, you are going to meet problems. *Who will own up to failing the company? Won't the finger of blame be pointed at others, by everyone?* What if people were simply doing their jobs, and the instructions they had been given, from on high, made little sense, so that the reasons for failure were actually seeded in policies, procedure, and processes?

*As difficult as it is to consult with people in order to get at facts, opinions, and the truth, it is a nonnegotiable element in building a system description. There is no way people will take ownership of a systemigram, the diagrammatic system of what should take place, if they do not find in it some of their own words. The beauty of a systemigram is that in the systemigram people can see themselves and their own words to some extent, via roles and responsibilities.* But most crucially they begin to see the roles of others and how the interplay between these various roles can be a source of confusion, illogical conflict, and competition as opposed to intended and desired cooperation. At least in this way the problem with the system is not the parts but the relationships. Once this is decided it is a matter of honesty and humility on behalf of each part to take (at least part) ownership of these relationships. It is this primary belief in integration as an ideal that drives the use of systemigrams and provides the leverage for achieving communion among a diverse group.

The process of gathering information and poring over relevant documents is highly labor-intensive. Skill is required not only to analyze but also to discriminate and to synthesize. Finally, artistry is needed to create a concise system description from all that has been garnered and processed, and thereafter a portrayal of that system description in the form of a systemigram. When all that is complete, one has to be prepared not for accolades but for remarks that range from "this is all obvious" all the way to "this is not how we actually do it." Each of these negatives warrants its own repost. To the former, it is, "Well it wasn't obvious at the outset!" And to the latter, it is, "Well, show us the way you actually do things."

The negatives do have an upside, which is to deflate the ego of whoever creates the systemigram. However, it remains the case that no comparable picture exists, in the form of a system, that captures the congregation of workers, the flow of information, and the logic to be efficient and effective as the company intends. The systemigram does this and in ways that are digestible, analyzable, and linguistically recognizable. And its utility is to vitalize the debate among all relevant stakeholders as to what is the nature of a problem, why it exists, how it came about, when and where it shows up, and who can do anything about it.

And such is what transpired in GEC Marconi. As you might expect, there were problems with the logic of some policies and procedures. This transferred into flawed processes that were fractured and dysfunctional. As people followed them, which they knew best to do, the consequence could be no other than failure. But there were other failings. An intense myopia, born of a need to focus, to be industrious and conscientious, or simply to keep one's head down, was widespread. Folk knew what they had to do; many had no clue what others were doing; and precious few had some notion of internal suppliers and customers, that is, the coworkers whose labors either benefited or were benefited by yours. In other words, *not even the meaning of "big picture" existed in people's minds. The focus for almost all was, "What's on the inside of my world?" and not "What is going on outside my world?"* The inclusion of the customer in a company process is a breakthrough, another grain of sand on which minds may turn toward new patterns of thinking and new ways of working that will benefit them and their jobs.

As a consequence of this effort, we are pleased to report that GEC Marconi, a company widely acclaimed for its engineering expertise, gave serious attention to improving its programmatics and reforming its culture. Systemigrams that captured end-to-end business processes were used to improve policies and procedures and as a basis for wide-scale courseware to support the training of middle-ranking engineers in preparation for future technical and managerial leadership roles. With that accomplished, systemigrams can safely disappear from view with more detailed definitions of best practice taking their place. *It is not important that systemigrams persist in an embodied form. It matters more that they live on in people's minds.*

# CHAPTER 17

# WHEN . . . ?

## WHEN IS THE TIME (FOR ANYTHING)?

At one time, we were huge fans of John Grisham. Having read one of his books, we could hardly wait for his next opus, and usually that didn't take too long. Over the past 15 years or so, the guy has been incredibly prolific, coming out with a new book annually, with many of them being turned into movies, at least in the early years. It is quite remarkable how this worldwide best-selling novelist has refocused a sharp legal mind from running his own law firm into creating gripping legal stories, works of fiction that bear a disarming similarity to current realities.

To be honest, it was the movie *The Firm*, starring Tom Cruise, Gene Hackman, and Jeanne Tripplehorn, that first got our attention. You know what happens: you see a movie, you like it, you think you might read the book. So you do, and then you want to read more so that you're "in the know" by the time the movie

*Systemic Thinking: Building Maps for Worlds of Systems*, First Edition.
John Boardman and Brian Sauser.
© 2013 John Wiley & Sons, Inc., Published 2013 by John Wiley & Sons, Inc.

comes out. You know you've found a pearl (of an author) when stars like Julia Roberts, Denzel Washington, Susan Sarandon, Tommy Lee Jones, and Matt Damon appear in the film of the book. This was the compelling attraction of Mr. Grisham, that his writing could assemble such a constellation of box-office power.

Our consumption of Mr. Grisham's output has waned in recent years; perhaps it is our general lack of interest in reading works of fiction as opposed to our current affection for the elevated literary skills of David McCullough, Joseph Ellis, and Ron Chernow that help us rediscover the majesty of our founding fathers. But one thing we have to say about Mr. Grisham that is of note is that we found his first book *A Time to Kill* by far one of his best works, notwithstanding the many he has penned over the last two decades. For an initial outing into fiction from an accomplished criminal attorney, it really could not have been better.

That particular book tells the story of an African-American father, Carl Lee Hailey, placed on trial for the murder of two men who raped his 10-year old daughter when it was clear as day to him that in the racially prejudiced South of the time, justice would not be meted out in response to their crime. So Carl, played in the movie by the outstanding Samuel Lee Jackson, takes matters into his own hands. Jake Brigance, played sensitively by an upcoming Matthew McConaughey, is the father's defense lawyer. Jake must argue the case that there is indeed "a time to kill." We recommend the book. The movie is okay, but the ending is the best bit by far.

Mr. Grisham must know his Bible, for the title of his first novel is taken from one of its books (Ecclesiastes). The relevant chapter runs along these lines:

> *There is a time for everything, and a season*
> *for every activity under heaven:*
> *a time to be born and a time to die,*
> *a time to plant and a time to uproot,*
> *a time to kill and a time to heal,*
> *a time to tear down and a time to build,*
> *a time to weep and a time to laugh,*
> *a time to mourn and a time to dance,*

> *a time to scatter stones and a time to gather them,*
> *a time to embrace and a time to refrain,*
> *a time to search and a time to give up,*
> *a time to keep and a time to throw away,*
> *a time to tear and a time to mend,*
> *a time to be silent and a time to speak,*
> *a time to love and a time to hate,*
> *a time for war and a time for peace.*

We are suitably emboldened by holy writ. There must be, we believe, a time for a systemigram, and therefore a proper time to create and to use one. Knowing exactly when that is what this chapter is all about. Our hope is to pass on our best experience.

## WHEN SHOULD YOU CREATE A SYSTEMIGRAM?

We teach many people, including large numbers of graduate students, *how* to create systemigrams. There is a huge and growing demand to learn this skill because *systemigrams are proving to be an excellent medium for communicating complex issues clearly and simply without being trivial.* Naturally, in our courses, we have to explain *what* they are and *why* they exist. But it has become very apparent to us that to overlook *when* they should be created and used would be an egregious error on our part. *There is a time to embrace (systemigrams) and a time to refrain.* We have come to recognize what makes it the right time by having looked back with care at over two decades of use in order to draw out the key characteristics of any given situation that make it fairly obvious that we should embrace and not refrain (and vice versa). So let's take a second look, here and now, at the circumstances that undergirded each of the three systemigrams we've presented thus far: the EEC ATMOSPHERE RTD project, the economic architecture of the U.K. Rail Industry, and the GEC Marconi Bid/No Bid process.

## Lost in Space

ATMOSPHERE had lost its bearings. A large number of talented people were behaving like sheep without a shepherd. In fact, although the project had a clearly identified management structure with Project Director, Technical Director, Marketing Director, Administrative Director, and so on, the voices that the sheep (the workers) heard were far from being in harmony; the sound was cacophonous. We soon came to understand that there was subsurface competition to give a technical lead from the principal representatives of the larger corporations, folks who felt they knew best and were perhaps under the gun from the seniors in their respective companies to deliver value for money, even though those applying this pressure might not have quite appreciated what the project was actually all about. In fact, no one in the ATMOSPHERE setup really knew what the project was about since the person who had conceived it and given it its original technical vision had moved on to new pastures and his legacy enshrined in the proposal document was sadly being ignored. This internal friction among the project's seniors generated much heat but little light, expended much effort—paid for by the EEC—but achieved little progress. The ensuing confrontation between the project's paymasters and its leaders was inevitable, but must it all end in tears?

Our reflections on ATMOSPHERE lead us to these conclusions. A large number of well-intentioned and talented individuals produced, for whatever reason, a whole that was considerably less than the sum of its parts. The project was failing in a great many respects yet the individual pieces of work seemed to be satisfying to those engaged in them. The project's leadership was aware of the considerable technical challenges of creating an integration environment (for developing software products). But it could not see how to address those challenges *architecturally*. There was so much emphasis on building something, to provide clear evidence of progress, as opposed to deliberating on abstract issues that might somehow be viewed as having little value. The problem came when what was being built had to be thrown away because nothing ever fitted together. We would say that ATMOSPHERE's failure was *systemic*, and this conclusion aligns with the complete lack of appreciation of thinking about systems per se as opposed to the evident expertise for building a system (but without reference to its context).

*The systemigram we provided, for the benefit of the entire project team, reminded all of the original technical vision, showed the key architectural elements needed to realize that vision, and provided the basis for comprehensive rational discussion among all of the varied technical perspectives as to what the challenges would be,* how they might (or might not) be overcome, and what collective executive action could be taken to make full speed toward an alternative agreeable overall goal. It served its purpose at a time when everything else was failing, and with all due deference to the esteemed Gene Kranz, failure seemed the only feasible option.

## No Way to Run a Railway

In the words of Gerald Corbett writing of the U.K. rail industry in the late 1990s:

> Privatization has delivered what its architects intended. But with success has come a problem—poor punctuality. Railtrack, responding to its incentive regime, has cut delays caused by tracks and signals but the Train Operating Companies, with some exceptions, have not made similar progress. This is because their economic regime is potentially lethal for punctuality. Trains are added, jobs are cut, punctuality pressures mount, but the growth hides any loss in fares due to the poor performance. But the problem will not go away. As the subsidies fall, the pressure will build and a recession would intensify it. Some train operating companies will be unable to keep investors and customers happy. The economic architecture will continue to drive them to actions that cause delays.

Here is a classic case of one principal stakeholder, within an extended enterprise, saying that the stakeholders *as a whole* are engaged in mission impossible, that no matter what any individual stakeholder did it would only make matters worse for the industry *as a whole* and that any short-term gain by these piecemeal actions would lead only to greater loss as the industry *as a whole* fell into further disrepair. This is systemic failure writ large. But consider the self-evident responses: "He would say that, wouldn't he? He has a vested interest" or "There he goes again, criticizing others while telling everybody the good his own company has been doing" or "He is saying that we, the government, got privatization wrong. It's easy to point the finger at us in order that he and the rest (of the profiteers) can hide their individual corporate failings behind this shameful cloak of blame."

These and many similar responses are all too obvious, aren't they? Obvious, commonplace, and devastating. What is anyone to do? Well before we try yet another approach let us look at the systemic implications of the words of Mr. Corbett. The *system as a whole* is made up of many private companies, each motivated to reward its shareholders. Each is incentivized to perform better, and these incentives are designed into the economic architecture of the *system as a whole*. Railtrack, who provide the bandwidth, is penalized for delays but is not rewarded for reducing delays. The train operating companies are rewarded for using the bandwidth, by virtue of increasing traffic, but are not formally penalized for delays. Should delays increase, their natural "penalty," or so it was thought, would be loss of traffic (with ridership migrating to their automobiles). In this case, the *system as a whole* is considered self-regulating.

Mr. Corbett, however, points out that the governance conditions for the *system as a whole* must inevitably lead to a demise of the *system as a whole* since its individual components will seek personal good, as they must, over collective good, this being a service to customers that is reliable and punctual (in addition to being relatively inexpensive and safe). He is asking these system elements to look at themselves not only as *wholes* but also as *parts* belonging to *a greater whole* and by so doing seek the good of the greater whole, which in turn will lead to the good of individual wholes that each naturally seeks. He is saying, "Let us not be myopic or parochial but let us see the big picture that we ourselves make up and that needs our attention, otherwise we won't be able to look after ourselves." We admire this. Our challenge is to build on that admiration by enabling this message to be shared among stakeholders, owned by the collective and used as a basis for concerted executive action. This is the only remedy for systemic failure.

*The systemigram we provided, for the benefit of the extended enterprise, illustrated the interrelationships between its key elements—the owners and operators of the newly privatized industry (with a place for the government's regulator)—together with the money flows between these elements, thereby exhibiting the economic architecture of the newly formed industry.* Each individual whole could thereby see how it was connected to others, operationally and economically, and thereby be able to explore the ramifications of the incentives and the impact of pursuing individual good at the expense of an otherwise imperceptible yet insidious deterioration in the service that the entire extended enterprise was to provide. Thus, as in the case of ATMOSPHERE, we were able to provide the basis for comprehensive rational discussion among all of the stakeholders as to what challenges they faced, how they might be overcome, and what collective executive action could be agreed, in this case to make full speed toward an efficient service to the traveling public.

## To Bid or Not to Bid?

Our work for GEC Marconi was the U.K. rail industry in microcosm. In the latter, the wider system represents an extended enterprise and its component systems are individual corporations. In the case of the former, the component systems are the functional units within the GEC Marconi company, and the wider system is in fact a key customer-facing process, the bid/no bid decision process, one that cuts across functional areas. For the systems thinker, the names in the frames have changed, but the basic architecture pattern, that of enterprise integration, is essentially the same. The integration mechanism for the latter was money flow and wider system incentives for the component systems. For the former, the bid/no bid decision process, integration was all about functional roles and responsibilities and the flow of information (and artifacts) that tied all of these together.

You would think that it might indeed be difficult to tie together autonomous corporations, regardless of the fact that they are all working for the same goal, for example, spectacular growth in rail traffic without compromising safety. And an extended enterprise is

for sure a tricky animal with myriad opportunities for misinformation, misunderstanding, and mistrust. In the particular case of the
Deepwater Horizon catastrophe, President Obama, in alluding
to the congressional hearings to which representatives from BP,
Transocean, and Halliburton were summoned, described one atro-
cious scene of pointing the finger of blame "a ridiculous spectacle."
The system had indeed failed, and we understand that the system
to which President Obama refers is in fact not the rig, though of
course that did fail, but rather the extended enterprise that drilled
and spilled.

You might further think that any given autonomous corporation
would certainly have its act together, that its departments were
experts in its particular functional areas, and that the corporation
as a whole was an integrated enterprise. Not so. At least not always
so. The emphasis on functionalism, an indisputable necessity, makes
the *integration of customer-facing processes, also a necessity, more
complex than it might appear*. There is a saying that "strong fences
make for good neighbors." That may be so. While clear departmen-
tal boundaries undoubtedly reinforce excellence of function, they
may also debilitate cooperation across functions. Departments may
be blissfully unaware of their neighbors and of the necessary col-
laborations between them that work on behalf of customers. Strong
fences may also increase a propensity for turf wars and other mind-
sets that create stove-pipes and ultimately institutional paralysis.
If no one is in charge of a nominated process, then it might as
well not exist. And if someone is put in charge, one of his or her
first challenges is identifying the course of action as a precursor to
improving and expediting flow. Once that landmark is achieved, the
turf wars flare up. At least with an autonomous corporation, it is
possible to "enforce" sound logic to make processes operate better.
That enforcement becomes more like an embrace, by the relevant
departments, once each of these components of the enterprise
system agrees with the bigger picture.

The systemigram we created for the bid/no bid decision (and
others that captured end-to-end business processes) provided that
bigger picture that previously had not existed other than as incom-
plete mental models in the minds of a few. This collection was
used to help redefine policies and procedures and to develop new

training materials to help the future leaders in the company to reorientate their thinking toward customer-facing business processes without losing the peculiar expertise in functional areas.

## WHEN WAS IT TIME TO USE *THIS* SYSTEMIGRAM?

### Dates and Dots

*"A date which will live in infamy."* With these words, the 32nd President of the United States, Franklin Delano Roosevelt, asked Congress as Commander in Chief for a declaration of war following the sudden and deliberate attack by naval and air forces of the Empire of Japan on Pearl Harbor, Hawaii, on December 7, 1941. He continued to say, "always will our whole nation remember the character of the onslaught against us." What is this character, exactly?

While the Japanese government deliberately sought to deceive the United States by false statements and expressions of hope for continued peace, they planned and executed a series of attacks across the Pacific, causing huge surprise among the U.S. military and the loss of many American lives. How could this be? Why the surprise? What was the state of intelligence on Japan's buildup of military might and of their planning to do great harm and pose grave danger to America? Where did it all go wrong? Who was responsible? When could it all get fixed?

William Joseph Donovan was born on the first day of 1883 in Buffalo, New York. His place of birth might have hinted at a different nickname, but maybe his Irish decent tipped the balance in favor of the one that stuck from the football fields playing for Columbia University and was reinforced by a distinguished service as an American soldier in World War I. Whatever it was, "Wild Bill" became an even more apposite epithet when Donovan was asked to lead the Office of Strategic Services (OSS), initially the Office of Coordinator of Information (OCI).

President Roosevelt invited Donovan in July 1941 to head up the OCI in order to overcome the fragmented operations of the various American intelligence services within the military and the

FBI. There was an uneasy feeling that while each service was capable enough in what it did, the real intelligence lay in what the vernacular of the day describes as "joining up the dots." The problem with representing autonomous enterprises as dots is they demonstrate a remarkable reluctance to be joined up, as we have seen so clearly in the case of the privatized U.K. rail industry and, on a micro scale, the functional areas of a single company.

The attacks on Pearl Harbor confirmed the worst fears that uncoordinated intelligence operations had proved an Achilles' heel in American defenses. Such was never to happen again. The Japanese attacks lent impetus to the transformation of the OCI into the OSS, and, after World War II ended, President Harry Truman signed the National Security Act of 1947, creating the Central Intelligence Agency (CIA), which took over the operations of the OSS. *The dots were now joined up and the graph was the property of the CIA. On paper.*

The sky was a peerless blue and entirely cloudless. Americans were at work on the East Coast and headed to work on the West Coast. It was just another beautiful late summer's day with business as usual. Planes were taking off, carrying their passengers on vacation, back home, and on business travel. America's economy was in full swing—but about to hit a bump. *Nineteen foreign nationals were on board four separate Boeings, fully loaded with fuel to carry them safely from east to west, from Boston and Newark to the Golden State, from sea to shining sea. Only one of these aircraft failed to hit its target*, later believed to be one of the symbols of political power in the nation's capital. The others had already toppled the icons of economic power and struck the Pentagon, symbol of American military supremacy. It was September 11, 2001. As time went by and people began to wrestle with the meaning of it all, the inevitable question was asked: "What happened to the dots?"

A symbolic successor to Wild Bill Donovan was identified. John Negroponte became the first Director of National Intelligence (DNI) and the new head of the 17-member United States Intelligence Community, replacing the Director of the CIA, who formerly had that role and who was now to "report" his agency's activities to the DNI. So now the dots are joined up again. No more dates to live in infamy. Not so far.

In response to the 9/11 attacks, a coalition of nations, led by the United States, struck the Taliban in Afghanistan. It was believed that the perpetrators of the 9/11 attacks had been trained there and that the organization behind the attacks, Al Qaeda, was safely harbored there planning its campaign of terror with relative impunity. Eighteen months later, the United States mounted Operation Iraqi Freedom. This had the deliberate goal of removing that nation's ruler, Saddam Hussein, who was suspected, according to U.S. intelligence reports, of developing, storing, and imminently about to distribute to international terrorist groups weapons of mass destruction (WMD). A rapid victory ensued and "mission accomplished" was declared in May 2003. To this day, no WMD have been located.

The Commission on the Intelligence Capabilities of the United States Regarding Weapons of Mass Destruction was created by Executive Order 13328, signed by U.S. President George W. Bush in February 2004. The impetus for the Commission lay with a public controversy prompted by statements that the Intelligence Community had grossly erred in judging that Iraq had been developing WMD before the March 2003 start of Operation Iraqi Freedom. President Bush gave the Commission a broad mandate not only to look into any errors behind the Iraq intelligence, but also to look into intelligence on WMD programs in Afghanistan and Libya, as well as to examine the capabilities of the Intelligence Community to address the problem of WMD proliferation and "related threats." The Commission, following intense study of the Intelligence Community, delivered its report to the President on March 31, 2005. We felt it was time for a systemigram, so we studied the Commission's report (aka the Robb–Silberman report) and created the systemigram that addresses the transformation required to make the U.S. Intelligence Community more capable.

## Transformation

The Intelligence Community (IC) has a primary responsibility to protect the United States and its allies. Easy to say, but how did the IC go about achieving this in terms of its prewar judgments about Iraq's WMD? Well, according to the cover letter that accompanied the report, by being "dead wrong"! We chose to illustrate this complete dereliction of duty in the bottom left quadrant of our systemigram, having laid out the overarching responsibility by contrast in the top right quadrant (Systemigram 17.1).

The report says the IC comprises "badly equipped and badly organized" agencies that use traditional techniques that have a declining utility against increasingly elusive and diffuse threats that seek to destroy the United States and its allies. These agencies include an "assuming analytical community" too slow and deficient in communicating to policymakers who help safeguard the United States and its allies. The IC, faced with threats from Al Qaeda, has an acute lack of human intelligence. This is perhaps understandable but is no reason for analysts to compensate for a genuine lack of intelligence information by making assumptions that are likely to take time to conjure and in some strange sense validate. In the end, telling people what you *know* to be true, as opposed to what you *assume* to be true, without telling them they are receiving the latter and NOT the former *is* dead wrong.

Additionally, the report continues, the IC suffers from "institutional incapacitation" (its phrase, not ours), which is created by "stove-piped intelligence expertise." This is what further frustrates policymakers, because what is known is not "joined up." There further exist two compounding errors: the assuming analytical community is likely to go undetected by the stove-piped culture, which also creates intelligence gaps and uncertainties that impede decision making by policymakers.

Hence the transformation program. This has four key elements: mission focus, Human Resources (HR) transformation (both of these to combat the institutional incapacitation), integration leadership, and information accessibility, the last being most needful to address the intelligence gaps (Systemigram 17.2).

Mission focus is akin to process focus with its intention to cut across (but not cut out) functional specialisms. The intended mechanism for achieving this is referred to as "target development boards," which have the aim of producing an integrated end-to-end collection enterprise in support of integrated intelligence (the primary goal of the transformation). Likewise, HR transformation will enhance the attraction of talent (to include extending and

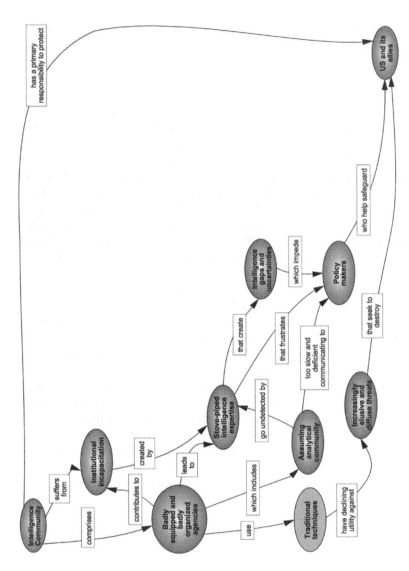

**Systemigram 17.1.** Intelligence Community: "Systemic Failure"

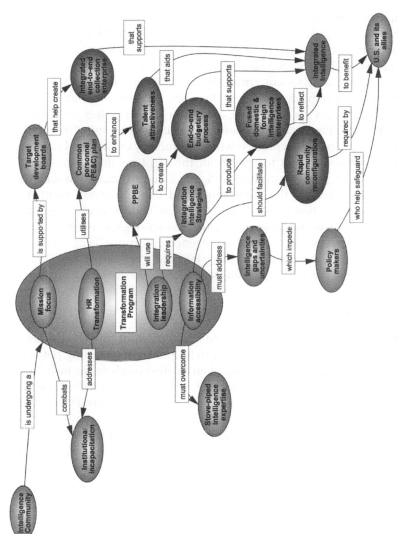

**Systemigram 17.2.** Intelligence Community: "Community Transformation"

augmenting human intelligence, especially collectors). To achieve this, the report proposes a common personnel (PE&C) plan across the entire IC. Finally, the report proposes implementation of an end-to-end budgetary process that supports the end-to-end collection and analysis enterprise by establishing priorities from an "IC-wide" perspective. These integrated priorities will enable IC decisionmakers to allocate resources to IC priorities through a disciplined Planning, Programming, Budgeting, and Execution (PPBE) process. This process will continuously evaluate all current and proposed efforts to ensure all the right needs get met (and efforts that no longer produce sufficient value are ended) to achieve the best effect for the community and the nation—in short, to achieve integration leadership. Integration leadership also requires well-developed "integration intelligence strategies" to produce a "fused domestic and foreign intelligence enterprise." These strategies would be articulated in the "planning" part of the aforementioned PPBE process (Systemigram 17.3).

## Integrate Issues

What are the American people to make of all of this? Is it a case, as appears to some, of persistent delinquency, a repeating pattern of unpreparedness? Of course, even professionals make mistakes, and the enemy is indeed becoming increasingly innovative, agile, and elusive. America and her allies face a sworn adversary resolute in their downfall and in the supremacy of an Islamic faith that is global in its domination, total in the subjugation of its own followers, and irrevocably dedicated to the extinction of the infidel. This is an enemy such that America (and its allies) has never known. Surely, this enemy is enough? America does not need *an enemy within*. The American people can tolerate neither incompetency among those who defend the country nor infighting among those sworn to protect it. And yet there is much evidence for the prosecution to make in the case for infighting and insularity. That is one side of the coin.

On the other side, from a practical perspective, it is no small matter to "join up" intelligence. First, the Community must serve two broad sets of customers—policymakers and operational

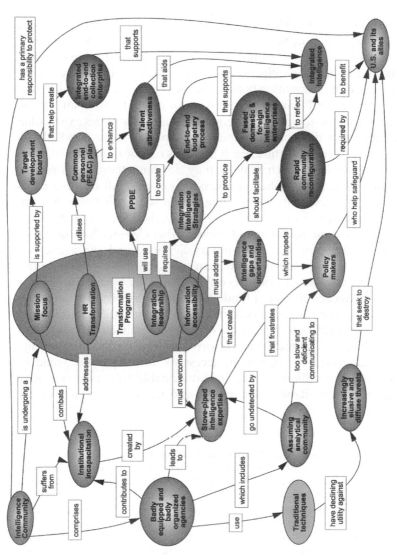

**Systemigram 17.3.** Intelligence Community: "Big Picture"

decision makers—who have distinctly different needs and often distinctly different planning horizons. Hence the IC's bifurcation into two "programs"—the National Intelligence Program (NIP)[1] and the Military Intelligence Program (MIP).[2] While there can be cross flow of information between these programs, it takes different capabilities, and many different *types* of capabilities, to attempt to satisfy each customer's requirements. These capabilities exist within a number of different intelligence disciplines, some of which are highly technical in nature. Each provides one or more sets of products or services, which may or may not lend themselves readily to integration with other products or services. All of these capabilities must compete for funding within whatever limits and priorities set by Congress, the White House, the Director of National Intelligence and Secretary of Defense, other Cabinet-level decision-makers, the Chiefs of the Military Services, and the agency directors. Because each has a different scope of interest and authority, at the practical level, priorities differ. Moreover, because of the way the intelligence budget is funded, the directors of the IC agencies can reallocate resources (within limits) internally. And because of the Community's real need for compartmentalization, creation of a "big picture" of intelligence efforts is an extraordinarily challenging endeavor; understandably, access to such a picture would be extraordinarily limited—and it is reasonable to expect that such access is not available to the people who are conducting day-to-day analyses.

[1]"The National Intelligence Program (NIP) funds intelligence activities in several Federal departments and the Central Intelligence Agency (CIA)." National Intelligence Program, accessed May 29, 2013, at http://www.whitehouse.gov/omb/factsheet_department_intelligence.

[2]"The term 'Military Intelligence Program' refers to programs, projects, or activities that support the Secretary of Defense's intelligence and counterintelligence, and related responsibilities as outlined in DoD Directive 5143.01. The term excludes capabilities, programs, projects, and activities in the NIP, and excludes intelligence activities that are associated with a weapons system whose primary mission is not intelligence." DoD Financial Management Regulation, Volume 2B, Chapter 17, July 2008, accessed May 29, 2013, at http://comptroller.defense.gov/fmr/archive/02barch/CHAPTER17.PDF, p. 16–2.

We recall our math teacher giving us a warning as we approached calculus for the first time, with its derivatives and integration techniques, that we were leaving behind the elementary things of mathematics and about to get into great (integrate) difficulty. His feeble pun left a memorable impression. Integration is difficult; it leads us indeed into great difficulty. The math is hard enough, but the issues, such as we have alluded to earlier, are infinitely more complex. Can systemigrams help? *Systemigrams solve nothing and they themselves are not a solution. But what they are is a pretty good example of integration.* We shall see how this is achieved in the next chapter. But we believe that to hold up a fine example of integration, or *systemicity* as some prefer, is to inspire others to emulate this in their own backyards, to encourage and support them in building excellent systems for themselves, regardless of complexity.

*Systemigrams, we argue, are part of a panoply of systemic media for problem owners. We use them to give ownership of a problem back to those who indeed own it and discourage it from being handed over to "problem solvers."* From our extensive travels with this medium, we have concluded that the time to use it is when problems refuse to go away, and solutions to those problems actually make the problem worse. It is also the time when many people are impacted by the problem, and many more become involved in problem-solving. This social firmament makes it more difficult to say what the problem is and more likely that solutions will be piecemeal at best and, what is worse, antagonistic to each other. The time for systemigrams is when change is less about policies, procedures, and processes and more about the people who conceive and use these artifacts, and how those people think; a time when no one wants to go first; a time when it seems people have time on their hands, a time to kill, when in reality the fierce urgency of now is calling them to attention.

# CHAPTER 18

# HOW . . . ?

A scouser (the local term for a native of Liverpool, England) is on holiday in Arizona, USA. He's staying in a remote, frontier-type town and walks into a bar. He orders his drink and sits down at the bar when he notices a Native American dressed in full regalia, feathered headdress, tomahawk, spear, the lot, sitting in the corner under a sign saying, "Ask me anything." The scouser is intrigued and asks the barman about him.

"Oh, we call him the Memory Man, He knows everything." says the barman.

"What do you mean he knows everything?" asks the scouser.

"Well, he knows every fact there is to know and he never, ever forgets anything."

"Yeah right," says the scouser, with reasonable skepticism.

*Systemic Thinking: Building Maps for Worlds of Systems*, First Edition.
John Boardman and Brian Sauser.
© 2013 John Wiley & Sons, Inc., Published 2013 by John Wiley & Sons, Inc.

"If you don't believe me, try him out. Ask him anything, and he'll know the answer."

"All right," says the Scouser, and walks up to the Memory Man.

"Where am I from ?"

"Knotty Ash, Liverpool, England," says the Indian. And he was right.

"All right," says the scouser, "that was easy, you probably recognized my accent. Who won the 1965 FA Cup Final?" (An annual knock-out soccer competition between the last two survivors in which hundreds of clubs enter. The Cup Final itself is normally held at Wembley Stadium, London, England.)

"Liverpool," says the Memory Man, quick as a flash.

"Yes, and who did they play?"

"Leeds United," comes the immediate reply from the Indian, without even blinking.

"And the score?"

"2-1," says the memory man, without hesitation.

"Pretty good, but I bet you don't know who scored the winning goal?"

"Ian St. John," says the Indian in an instant.

Flabbergasted, the tourist continues on his holiday and on his return to Birkenhead (a town just across the River Mersey from Liverpool) tells all and sundry about the amazing Memory Man. He just can't get him out of his mind and so he vows to return and find him again and pay him his due respect.

He saves his dole money (unemployment benefits) for years, and finally, 12 years later, he has saved enough and returns to the United States in search of the Memory Man. He searches high and low for him. And after 2 weeks of trying virtually every bar and town in Arizona, he finds him sitting in a cave in the mountains, older, grayer, and more wrinkled than before, but still resplendent in his war paint and full regalia. The scouser, duly humbled, approaches him and decides to greet him in the traditional manner.

"How."

The memory man squints at the scouser.

"Flying header in the six-yard box."

\*\*\*\*\*\*

And the point? Even a simple three-letter word holds multiple perspectives.

## HOW DO YOU CREATE A SYSTEMIGRAM?

Talk it up and write it down. The very first thing we have to say about how to create a systemigram is that it is *never* done without text. What is more, unless that text bears the hallmarks of being a system in its own right, there is no possibility that the systemigram designed on the basis of that text can be considered a system, and a systemigram that is not a system is not a systemigram. The question that arises therefore is how does one create the textual background whence the systemigram is created, and in a fashion that makes that description a system in its own right?

It is important to remember that the textual description is actually a description—it describes something, and that which it describes must be worthy of note, and not trivial. Its worthiness is a reflection of the issues it enfolds and the various perspectives or viewpoints of the differing interests that affect and are affected by those issues. Usually, this mix is a complex, an interweaving of agendas, priorities, opinions, biases, and even cultures. The textual description is the first attempt to portray that complex in a manner that makes it relatively simple to understand and therefore accessible to more objective perspectives with which the complexity might be not only duly respected but also dutifully unraveled. Technically, we refer to the mix or complex as the System of Interest (SoI), and the textual description that initiates access to this as the SoI description. Whether the SoI is or is not a system, the SoI description must most certainly be, and hence must the systemigram.

What ensures the SoI description to be a system? You might as well ask, What ensures any example of good writing to be a system?

Take two contrasting examples of a speech, one very brief—the Gettysburg address—the other rather lengthy—the Inaugural address by the ninth President of the United States (who ironically had the shortest tenure—we assert no causation here!). How do you judge either one of these to be a system? Consider books. What makes *The Silence of the Lambs* a systemic work of fiction, or *Down Under* an unmissable travelogue from Bill Bryson? We have concluded that these last two examples (and most certainly the Gettysburg address) are indeed systems. In short, these works hang together; they have structure and they have dynamism, in the case of a good book symbolized by the accolade "page turner." The parts of the work bear significant relationships with one another, via invented characters and their dialog or scenic descriptions and their sensory impact on the reader. These parts and relationships form a coherent, harmonious whole. This whole has evident structure and discernible process (or dynamic). It has variety, for example, a wide vocabulary, and parsimony, an economy of expression, all of which is harmonized, resulting in an elegance and a beauty greatly appreciated by the reader who becomes absorbed in the message the work seeks to convey.

The task of creating a systemigram is made so much more straightforward if this preceding text, or SoI description, is fully deserving of being called a system. That task is rendered impossible if no such text or a scant description exists. It's not necessary for the creator of the systemigram to be the author of the SoI description, but it helps. When this is not the case, it behooves the systemigram designer to be critical in his or her judgment and evaluation of the SoI description, and to be so means to judge that the SoI is worthy of full description and the writing itself as systemic as can be. We are firm believers that good writing follows much reading, that if anyone seeks to be a good (even great) writer, he must first be an assiduous and eclectic reader. In that way, the creative forces at work are being properly exposed to multiple sources, continually subjected to good works and bad, to great works and to a deep appreciation of the skills, craft, and technique of observers, commentators, communicators, and messengers of all walks of life.

The SoI description contains a complex message that is made less complex by the forensic trail of linear text, and the systemigram, by virtue of its fundamental visual nature, makes visible both the complexity of that message *and* the simplicity of the text. In that sense, the purpose of the "how" relative to creating or designing a systemigram is to invite the problem owners associated with the SoI to respect the complexity of the system in which they are embroiled while committing to its resolution as a collective. The designer of the systemigram is obliged to leverage the basic architecture of this visual framework in order to exhibit complexity and simplicity *simultaneously* in a way that cautions against oversimplification yet exhorts to addressing complexity in a measured cooperative fashion.

## Rules and Principles

The principles that govern the architecture of a systemigram are identically those that govern inspection of any SoI using wellfounded systems concepts. Among these are boundary, relationships, transformations, function (or purpose), emergence, harmony, and control. With each one of these concepts, we can, via the notion of equilibrium, associate two more, thereby making seven sets of triples. We have spent time to package this collection of seven triples into a single artifact that we have used throughout our systems careers and that we believe is strongly influential in deploying the systemigram technique. This we have called the Conceptagon, and we made it the subject of our second journey.

Thus, when we browse the SoI description prior to designing a relevant systemigram, we cannot but notice, for example, parts, relationships, and wholes, and inputs, outputs, and transformations. It is the way we regard text when we are reading, at least from a mechanical viewpoint. But we cannot also fail to observe the forces of variety and parsimony at work to produce harmony, and indeed hierarchy, openness, and emergence. For us, this manner of systems thinking is how we achieve a fuller understanding of the message that the SoI conveys, and gives us the momentum to translate this into a systemic diagram.

With that said, we are now in a position to define some simple rules that help achieve this transformation. First, the nodes in a systemigram are always nouns or noun phrases. Naturally, not every noun in the SoI description appears as a node. Parsimony helps us choose only those entities that appear to be most significant. However, once a node, that noun phrase is unique in the systemigram and cannot be replicated. So therefore, everything that semantically attaches itself to that noun phrase in accordance with the text must be accommodated in that single appearance. However, not everything that could be said is chosen; only those most significant expressions and relationships with other noun phrases, so chosen as nodes, are present in the diagram. These relationships, represented as arrows, are the verbs and verb phrases (and sometimes prepositional phrases) from the SoI description. Attaching significance is a matter of sound judgment by the systemigram creator, aided and abetted by the sound writing of the author of the SoI description. The arrows must not cross one another. This rule helps maintain clarity while ensuring the correct degree of significance is accorded.

Some nodes can be made to include other nodes (in which case they are known as containment nodes). Arrows can enter and leave a containment node and/or any of the nodes within it, so long as consistency is maintained with the SoI description. This kind of convexity supports, for example, the key elements of a tender submitted to a customer, the weapons catalog used by an adversary, or the portfolio of an enterprise engaged in strategic intent. On occasion, an arrow can pass through a containment node, thereby circumventing a crossover that might otherwise occur when insisting upon an additional relationship that simply cannot be omitted.

Digestion of the SoI description usually leads the systemigram creator to an inescapable conclusion as to not only the message, but also its principal subject and objects. These two have a special place in systemigram geography; the former is placed in the top left corner and the latter in the bottom right corner, the flow of the message then proceeding from subject to object via numerous threads, which may include return flows. One particular thread, held to be the most significant, is the mainstay of the systemigram, and ordinarily will flow diagonally from subject to object. Commonly, this is the mission of the system that the SoI description harkens to, but other notions are permissible so long as the mainstay captures the central theme of the SoI description.

Systemigram geography can be important in serving the other needs of generic themes to which the SoI description will often allude. For example, every mission has a motivation and a management or governance mechanism. Missions are fraught with obstacles and plagued by failure. Missions are renewed by insightful strategies and their orderly execution. All of these features— motivation, management, failure, and remedy—can find a proper portion of the systemigram itself. Intriguingly, noun phrases, the nodes, that belong to any of these may also belong to or be surprisingly connected into tense regions, a feature that satisfies the need for integration of ideas to which all systems aspire. This is a great opportunity for creativity by the systemigram designer and will often portray the added value of a diagram that linear text alone cannot provide. With that artistic creativity comes a deeper appreciation from the audience that receives the systemigram of the system with which they are faced, and that appreciation provides much needed momentum when elements of that audience must learn to collaborate, sometimes for the first time and often as former opponents.

## Mutate and Evolve

It is our experience that many people greatly admire a finished systemigram. That is gratifying. What people fail to appreciate and never see are the many failed attempts that preceded this completed work. Failures that led to eventual success but which at the time were nevertheless hopeless rejects that never deserved to see the light of day and which at the time made life for the systemigram creator pretty miserable. That said, it never ceases to amaze us how those very failures somehow contained the seeds of eventual success. It is amazing but true, and we are obliged to allude to this aspect of systemigram design in our description of "how." The simple lesson is: be prepared for many, a great many versions of what it is you are seeking to achieve to end up in the trash can. Somehow, each successor, though in all likelihood to become a reject, is an improvement on its predecessor. It's as if there is some strange mutation at work under the influence of a mysterious evolution process, one that fits the completed work for survival.

What are the signs to look for that mark "defects" in early life forms of the systemigram species? The need to make links cross over for the want of adding necessary significant relationships between nodes is one. Sometimes, this is easily fixed by simply moving nodes around. Sometimes, the fix is more exotic, calling for a deeper understanding of the SoI. These defects ought to be regarded not as problems to be fixed but rather as opportunities to better understand the message of the SoI description and hence to portray that message with greater dignity and beauty, scaling new heights of elegance in the systemigram design.

Another major source of defect is the multiplicity of start and/ or end points. By insisting on a single source and terminus, the systemigram is best able to clearly demonstrate the subject matter and its objective, together with all the flows of information, activity, and interdependencies that lie in between. Look again at these "stray" sources and termini. Are they *that* significant in the SoI description? Or are they even more significant, requiring a reevaluation of their portrayal and interconnectivity with other nodes?

One final defect we want to mention is the "apparent" completed work. It is exceedingly tough to reject what for all intents and purposes looks like a beautiful systemigram, one that holds together and conforms to all the rules of design. But as attractive as it might be, does it really tell the whole story? Does the geography of the systemigram reflect the major themes of the SoI description, which might be, for example, vulnerabilities, threats, security, and mission? These generic themes or others that are appropriate given a thorough comprehension of the SoI description should be clearly demarcated by the topology of the systemigram, and they should interrelate as the SoI description warrants. The *interdependency* of these themes will be reflected in the various links that tie nodes together, nodes that play a role in one or more themes. Getting these interdependencies correct while properly portraying the themes can call for a rework of what otherwise looks like the perfect picture. The lesson here is that the systemigram designer, while fully appreciating the values of elegance, beauty, and harmony, must also be committed to the truth, and the search for that, while not endless, often calls for us to go the second mile. Make the effort, it will be well worth it.

When does this evolution end, if ever? To answer that question, we turn to a new perspective on "how," one that enables us to consider a compelling exhibition of the systemigram itself.

## HOW DO YOU CREATE A SYSTEMISHOW?

### Like a Box of Chocolates?

We often ask our students if they recognize the name of a person we give them. In one such quiz, we ask, "Ever heard of Winston Groom?" Blank looks. "How about Eric Roth?" is a follow-up question. Blankness persists. "You must have heard of Robert Zemeckis," we assert. A few lights go on around the room and one or two smiles creep across the transformed faces of a few. Most heads are nodding by now, though we can't say for certain that this is not a social function influenced by the few who really seem to recognize this third person. We are now inclined to put people out of their misery. "What about Tom Hanks?" we ask. Unanimity among the crowd. No anonymity for Tom, unlike Winston and Eric. Dare we ask the class to join up the dots? "What's the connection between these four?" we ask.

Obviously, if you don't know the first two people, then joining up the dots is tricky. But maybe two dots are enough, and a few aficionados know the movies that Zemeckis has directed in which Hanks has starred. *Cast Away* is one, made in 2000. But 6 years previous to that movie came one that swept the board at the Oscars. Its name? *Forrest Gump.* The movie won Best Picture that year and gold statuettes also went to Hanks and Zemeckis. It is little known that Eric Roth won one for Best Adapted Screenplay. And the guy who misses out, the creator of Forrest Gump, whom we all know better as Tom Hanks? You've probably guessed: Winston Groom. And the point? It takes *a series of systems* to create a work of art that makes such widespread impact, bringing tears and laughter to millions.

First, there is the book; thank you, Winston. Then the screenplay; credit to Eric. The producers have to believe in the book and the character(s) to bring the story to development, and for *Forrest*

*Gump*, those guys get a reward on Oscar night. And the whole movie needs to be steered from screenplay to screen using a complex enterprise of cast and crew, among which are the various artistic talents of Garry Sinise, Robin Wright Penn, and most memorably, Tom Hanks, who unforgettably brings the eponymous hero to life and for our lasting enjoyment. Books and writing talent, production and producers, development and directors, screenplays and more writing talent, and finally actors, including stars, and their individual and collective performances. These are the systems that make us recognize Forrest Gump and what that character means. We might boil it all down to: "Life is like a box of chocolates, you never know what you're going to get." But the truth is that the whole meaning we derive from Forrest Gump is a tribute to that series of systems.

While not anywhere in the same league as the Hollywood blockbuster, we like to think of the systemigram as a collection of systems assembled together for the purpose of bringing insight, inspiration, and even entertainment to a crowd of interested observers. And why do we choose to regard a systemigram in this manner? Because our interest is in doing all we can, *systemically*, to help those with a direct involvement in the System of Interest to move forward. Our hope is that the insight and inspiration that a systemigram can bring will be leveraged so that the community of its observers (or metaphorically its theatrical audience) can take its responsibilities more seriously and more studiously given the greater context in which the systemigram places them, so it can provide real momentum for concerted action among the individual observers and move them forward as a community. With that said, we are now ready to talk about the systemigram as a series (or system) of systems culminating in an exhibition that seriously engages observers.

## Genesis of Systemic Media

What is it that kicks off this "collection of systems assembled together for the purpose of bringing insight, inspiration and even entertainment"? The very first system is the System of Interest. Let us take one specific example: improvised explosive devices

(IEDs). Any one of these devices could be a specific system of interest. Systems professionals could tell you how it was made, how the various pieces were obtained, how the device was employed, and so on. All of these interests apply to any specific IED. At another level, there is interest in how the improvisation process is conceived, resourced, and deployed. It is this process that leads to the realization, lethality, and ultimate effectiveness of any given IED. The interest level has changed, the system level has changed, and so the SoI has changed, but in some way or another, all of these perspectives apply to the IED phenomenon. How troops respond to this phenomenon is yet another SoI. It is easy to see how the picture and therefore the problem can become rather complex, and how confusion might arise among the interested observers of one or more of the relevant systems of interest. We maintain that systemic thinking is an appropriate response to this complexity, and what we shall call systemic media is our preferred and recommended systemic thinking approach for addressing and resolving this complexity. Systemic media allows us to progress swiftly though not hastily from the first system, the SoI, to the next one, the SoI description, a transition that calls for sound inquiry and analysis, but also a special kind of synthesis, namely, writing.

We would not presume to tell you how to write; it is enough for us to struggle with this complex phenomenon. What we do say is that we firmly believe that in order for us to become better writers, we need to be bigger readers. It really doesn't matter much what the reading material is, as long as it is good writing! Fiction, biography, political history, economic commentary, as well as popular and mainstream science and technology are all to be enjoyed. If sound writing is read with a systemic thinking cap on, so much the better, provided of course this does not spoil your enjoyment of the subject matter.

What we would also say is that many systems of interest are in better shape than the SoI description that seeks to capture them. This perhaps betokens a proclivity in many, maybe most, for action rather than contemplation. Notwithstanding Jacob Bronowski's assertion that "the world can only be grasped by action, not contemplation," the two are not incompatible. What is more, the former benefits greatly from the excellence and execution of the latter. Albert Einstein said, "In the brain, thinking is doing." The two are essential parts of a greater system, and their relationship should be safeguarded and nurtured. And so we argue that if the SoI itself is the first system in our collection of systems, then the second one is the SoI description, and to make this worthy of the moniker *system* requires appropriate writing skills.

In our work, using systemic media we have sometimes been disconnected from the SoI, for reasons of security, confidentiality, or other such issues. That disadvantage has not necessarily impeded our efforts because we have been given access to reliable SoI descriptions. This can bring us into direct contact with their authors who have direct involvement in the SoI. In any event, we needed to compile our inventory of problem owners of the SoI. Sometimes, the SoI description is lacking (or nonexistent), and our work can help remedy this by providing an even better understanding of the SoI for the problem owners than the descriptions available could do. Sometimes, we wondered why, as in the case of ATMOSPHERE, the SoI was performing so poorly when the SoI description could not have been more lucid and comprehensive. Treating these two systems separately but interdependently is key.

If the SoI description is the "book," then the next and third system in the series, the systemigram, is the screenplay. Recall that while the systemigram is a diagram, it is also a system, and one aspect that makes it a system (more than just a diagram itself being a system) is the fact that it is created from the SoI description. It is a diagram based on the "book." The diagrammatic structure metaphorically is an analysis of the cast, the individual characters, the dialog and actions that take place within and between characters, and the themes that surround the SoI, hopefully expressed and emphasized in the SoI description itself, which in an intriguing sense become the plot lines, subplots, and tensions that fill and make any story.

It is of course true that many screenplays are original works and not adaptations of a book. But these screenplays are textual, often calling for excellent writing skills. They are also logically consistent with the message or story that makes them vital and so in our world are true to the SoI. And screenplays are often rewritten, with great care, as a consequence of the vision of producers, crew, and cast. Movies are about words and pictures. They are also about structure and process, being and doing. Movies have both composition and dynamism. They afford an excellent metaphor for our systemic media. The SystemiShow, our fourth and final system in the series, is as near as we can get it without having the production values of an *Avatar*, the movie based on the screenplay or systemigram. It is this that goes on exhibition to all those with an involvement in the SoI. Will it be a blockbuster, will it "make money," will it stir (and turn) hearts and minds? These things we cannot foretell. What we believe is that systemic media is the best shot that systems professionals have of helping problem owners who wrestle with the contemporary complexities of our world today, the impact of which escapes very few in this age of systemic failure.

### Hints and Tips

Knowing that the SoI description can safely be regarded as a system and that well-developed systems thinking has governed the translation of the SoI description into a systemigram, then the creation of the SystemiShow—this being the theatrical exhibition around which the SoI problem owners eagerly gather—can also be regarded as a system and can take many cues for its design from the themes, messages, plots, dramas, characters, and the actions they make and the lines they speak, and the overall intent that the SoI description undoubtedly possesses.

We believe that it is essential to energize the problem owners with what each knows they are about but also about all the other things that each may not know about their neighbors. It is key to bring them all together and have them confront the reality of the community they constitute and the responsibilities expected of that community by outsiders with less than a direct involvement, certainly as far as the governance of that community is concerned. The systemigram will do that, as will the SystemiShow. But what the latter can do additionally is help the audience better understand the complexity of the SoI and appreciate how autonomous action by individuals, while predicated on the interest of others as well as self, can actually lead to counterproductive consequences. It is only when this autonomous action is guided by a full appreciation for context that it might serve as intended. That full appreciation can be effectively communicated by a compelling SystemiShow.

Over the years of working with this medium, we have learned much ourselves, and in this brief section, we want to pass on this learning with the specific intention of encouraging showmanship in two key senses: the traditional one of putting on a show, and the less conventional one of ensuring that scholarship and scientific method are not betrayed for the sake of the former. It is not easy to hold in balance these two. The tendency is for one to predominate, leading to a glitzy but insubstantial exhibition or a thorough yet uninspiring accounting. The purpose of systemic media is to enable systems experts to exhibit to problem owners their genuine professionalism, one that will lead to an admiration of systems qualities such as elegance, beauty, harmony, truth, and context, which in turn will produce a desire for these to be found in the SoI. Once problem owners have that desire and it is accompanied by an invigorated regard for collective action, their preferred program of executive action, whatever that is, will be the beneficiary. Systemic media can become a distant memory; the SoI resumes rightful primacy.

Here then are a few hints and tips. First, it is good to *show off the characters*. In the systemigram we shared in Chapter 15, these would include the train operating companies, Railtrack, and the government regulator, OPRAF. Each has roles and responsibilities, but each certainly has a distinct personality that leads to lines of action and to a style of communication particular to the personality of each. Some of this will have been captured in the SoI description, but much can be interpreted from a reading of the whole and an appreciation of the culture that prevails in that industry can be formed. These characters (often better known as stakeholders or problem owners) need to be given due prominence in the show so that one scene, for example, might show them naked, as it were, with no links evident, and supporting commentary might allude to battle lines needing to be drawn or barriers to be overcome.

Second, it is good to *demonstrate the dialog that takes place between characters*. This can be bilateral or multilateral or a thread of conversation composed of a series of bilateral exchanges. In Chapter 16, we shared a systemigram that captured one corporation's strategy for filtering opportunities to bid on contracts by investing in a bid/no bid decision process. The characters here are roles, such as bid proposals manager, commercial department, general manager, and review team. These roles in real life would be assumed and played out by actual people with appropriate skills and knowledge and of course certain attitudes toward others and their world of work. In this systemigram there is a lot of information flow and opportunity for dialog. Telling the story scene-by-scene can present an amazing opportunity for dramatizing workplace culture, which is another level up on "mere" process definition.

Recall that systemic media kicks in because there is "a problem" with the SoI. Sometimes, this is bad process definition. Often it isn't; the problem can be rooted in culture that influences communications, attitudes, and cooperation, matters that may go untouched by process reengineering. The SystemiShow designer should take every opportunity to reflect upon this duality—of what is supposed to be the case and why humans will find out. It is all part of the act of juggling and putting on a show while honoring scientific method.

Third, it is wise to *make each scene a digestible piece of information*. We have already commented previously on how a causal loop diagram describing counterinsurgency operations in Afghanistan came in for extensive criticism from the media and the military. People are curious. It is perfectly okay for them to use an iPhone and all its amazing services. They don't want to be confronted with the complexities of the A4 chip that enable these services. The mistake made with the causal loop diagram was to present it as though it was understood or understandable. This is a mistake often made by enthusiastic inventors, discoverers, and researchers, eager to share their results with those they foresee as being the beneficiaries of their work. Sadly, those latter people have not experienced the journey of discovery or the labor of invention. The SystemiShow is an opportunity for "the novice" to experience that journey culminating in the entire systemigram being remarkably understandable at the end when at the outset it would be dauntingly incomprehensible.

A fourth tip is a follow-up from the previous one, and it is to ensure each new scene benefits from the audience's understanding of the previous one and indeed all previous ones. In other words, *adopt an incremental approach to scene development that supports growing audience awareness and understanding.* Of course, this relies on good storytelling skills. The SoI description may be composed so as to support good story telling. But it may not. When the systemigram is designed, not only must its creator be faithful to the SoI description, but he or she must also have in mind the story that is going to be told later with that systemigram. That story telling can come about only through a deep understanding of the SoI description—including a knowledge of what it fails to say because of its writers' evident neglect—so that the systemigram itself adds value without putting an underserved spin on the SoI description.

It is important to point out that the audience can objectively criticize the SystemiShow (and systemigram), since its author(s) could have got matters wrong. This does not amount to failure, unless the exposure totally spoils it for systemic media. If the audience likes the attempt and can find clear fault lines, that is success. The creators of systemigram and SystemiShow (and possibly of the SoI description) are now their own community endeavoring to shine a better and greater light on the SoI. Reworking the media to serve this end is no hardship. In the end, remedying the SoI is really all that matters.

A fifth hint refers to the commentary that accompanies the SystemiShow exhibition. We need to remind you that the systemigram does not capture all that the SoI description contains, but it does get at the essence. Likewise, the SoI description does not capture all that the SoI contains, but it does get at the essence. By the same token, the SystemiShow cannot capture all that an audience needs to experience, but it should provide an essential platform to support additional commentary, explanation, and insight. When the commentator is indispensible to the SystemiShow, it has failed. We have experienced this many times both as teachers and as consultants. People have said, "Systemigrams are great when Brian (or John) present them, but without him, they don't work!" Commentary is essential, and it should be part of the SystemiShow even when the latter has ceased. Commentary is part of systemic media, but it must be a natural partner to SystemiShow. Commentary should not replace SystemiShow or confuse it, but it should bring it out into the light where it can happily continue long after the commentator has gone. Our advice is this: *make commentary a natural accompaniment to SystemiShow*, and if you are unable to do so, rework the SystemiShow so that your commentary becomes its servant and not its master.

And finally! When the audience sees something new in the systemigram that they got from the journey, that is, *from* the SystemiShow, accept that as a form of validation. Systemic media is in the value-adding business. Sometimes great value can be added, sometimes little or none. Judging what value systemic media can bring is part of the professionalism that comes with assembling the relevant tools. In the end, however, the SoI is what really matters. What systemic media can do is assemble the problem owners of the SoI, make them aware of the community they constitute, give them a new regard for the complexities of the SoI, and provide them with forward momentum to execute whatever remedies they can agree upon as a community, remedies that hopefully benefit from fresh insights and inspiration.

## SHOWTIME: THE IED PROBLEM: A SYSTEMIGRAM STORYBOARD

The IED problem is illustrated in the following systemigrams: Systemigram 18.1: Scene 1, a device mentality; Systemigram 18.2: Scene 2, traditional weapons of war; Systemigram 18.3: Scene 3, attacking the head; Systemigram 18.4: Scene 4, the insurgents' weapons; Systemigram 18.5: Scene 5, explosive kills; Systemigram 18.6: Scene 6, inverted criticality; Systemigram 18.7: Scene 7, a paradigm shift; Systemigram 18.8: Scene 8, modern weapons of war; and Systemigram 18.9: Scene 9, the big picture.

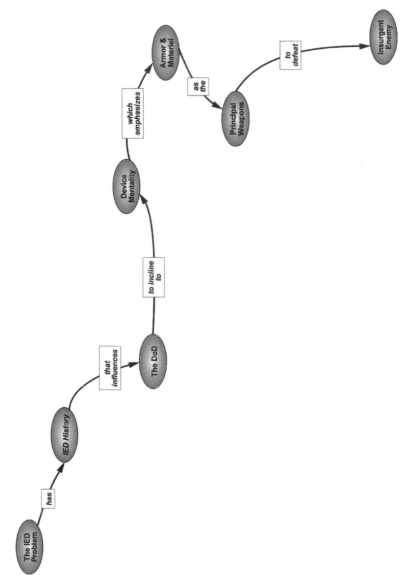

**Systemigram 18.1.** IED Problem—Scene 1: "A Device Mentality"

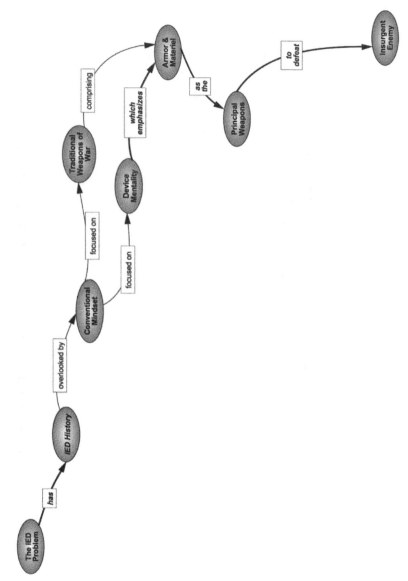

**Systemigram 18.2.** IED Problem—Scene 2: "Traditional Weapons of War"

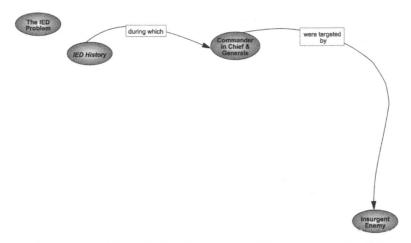

**Systemigram 18.3.** IED Problem—Scene 3: "Attacking the Head"

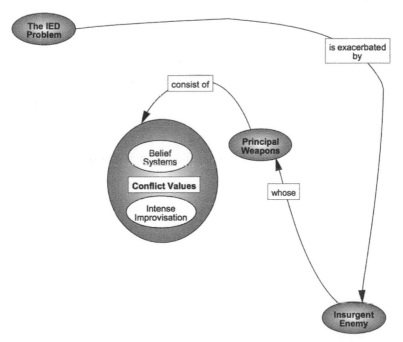

**Systemigram 18.4.** IED Problem—Scene 4: "The Insurgents' Weapons"

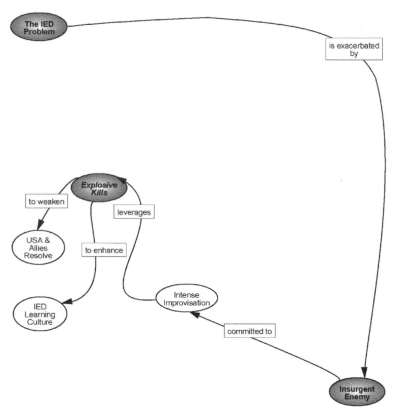

**Systemigram 18.5.** IED Problem—Scene 5: "Explosive Kills"

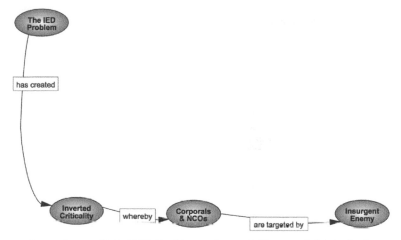

**Systemigram 18.6.**  IED Problem—Scene 6: "Inverted Criticality"

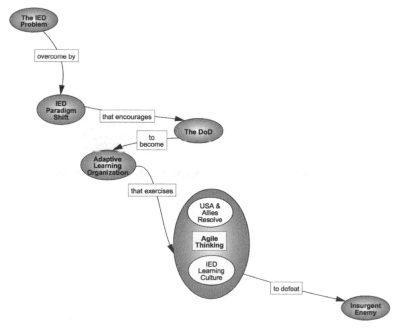

**Systemigram 18.7.**  IED Problem—Scene 7: "A Paradigm Shift"

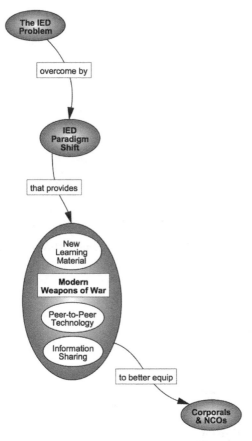

**Systemigram 18.8.** IED Problem—Scene 8: "Modern Weapons of War"

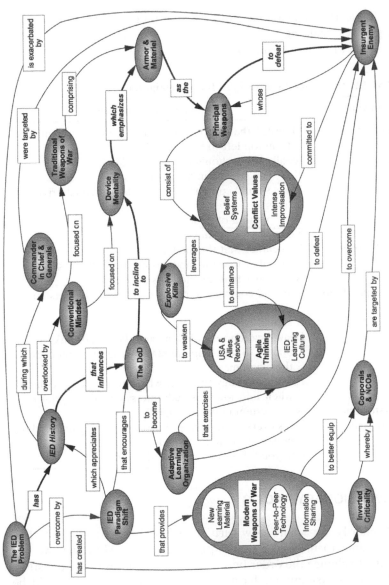

**Systemigram 18.9.** IED Problem—Scene 9: "The Big Picture"

Table 18.1 is a summary of systemigram guidance and the principles for building a model. This table can aid in validating the essence of your systemigram.

**TABLE 18.1. Guiding Principles for Systemigrams**

| Principle | Systemigram Guidance |
|---|---|
| Correctness | Mainstay that supports the purpose of the system reads from top left to bottom right. |
| | Ideally, there should be 15–25 nodes. |
| | Nodes must contain noun phrases. |
| | Links should contain verb phrases (to reduce trivial links). |
| | No repetition of nodes. |
| | No cross-over of links. |
| Relevancy | Remember that the model is really "theirs." |
| | Remember that the model is not really "theirs." |
| | Remember that the model is not reality. |
| Comparability | It should compare to reality and the original system description. |
| Clarity | It should read well. |
| | Beautification (e.g., shading and dashing of links and nodes) should help the reader read the sentences in the diagram. |
| | Exploit topology to depict why, how, and what (who, when, and where is built into system description). |
| Systematic design | Is it a system in its own right? |
| | Does every node (except for the beginning and ending nodes) have an input and an output? |
| | Can you follow any node to the end node? |

## DIVERSIONARY TACTIC

Ordinarily, we would now turn to "where" and finally to "who" according to the order that supports rhyming. However, we are going to change the order because "where" applies to the question, "Where is systemic media headed?" And that we feel should be left to the end. We offer our sincerest apologies to Mr. Rudyard Kipling, believing he would understand and concur.

# CHAPTER 19

# WHO . . . ?

Who has not seen at least one episode of *I Love Lucy*? Perhaps today that number is large and growing, but not many years ago, the answer may well have been a resounding "No one!" The show ran on our black and white TV sets during the first half of the 1950s. Reruns in the decades since, both in the United States and globally, have ensured a global reach for its female lead Lucille Ball, a zany redhead with a unique comic flair who also possessed an uncanny knack for business. Typical of such an iconic form of entertainment, clips of the show appear in blockbuster movies, such as *Crocodile Dundee II* and *Pretty Woman*. The latter movie launched the career of yet another beautiful redhead whose business savvy in knowing which parts to choose, when combined beautifully with her considerable acting ability, ensured an Academy award for Best Actress playing the eponymous heroine in *Erin Brockovich*.

Undoubtedly, more people today have heard of Julia Roberts than Lucille Ball. Yet the latter was an accomplished actress

*Systemic Thinking: Building Maps for Worlds of Systems*, First Edition.
John Boardman and Brian Sauser.
© 2013 John Wiley & Sons, Inc., Published 2013 by John Wiley & Sons, Inc.

appearing in many films throughout the 1930s and 1940s. Notably, in 1968, she played Helen North, a widow with eight children who falls in love and marries a widower. Now it's tough enough for a guy suddenly to find himself a father of eight. But the movie is even more intriguing than that. Frank Beardsley, played by Henry Fonda, already has 10 children of his own. The hapless newlyweds immediately find themselves as the parents of 18 children! Is it possible for all of them to come together as one big happy family? Can an agreeable integration be formed? Will they be a dysfunctional assortment or can they constitute a well-structured, dynamically stable system? Or will it be a case of "yours, mine, and ours" (which happens to be the movie's title)?

This meandering preamble has purpose. When it comes to addressing the matter of *who*, relative to the creation and deployment of systemigrams, what we want to say can be comfortably accommodated by the notions of *yours, mine,* and *ours*. This is important. To know clearly not only who is involved but who does what for whom and who owns what in the process is to fully appreciate the sense of systemic media.

We take these two viewpoints: the *consumers* of systemic media, aka the problem owners, is *you* (and what *you* own is *yours*), and the *producers* of systemic media is *me* (and what *me* owns is *mine*). Together, producers and consumers represent *us* and what *us* owns is *ours*). The purpose of systemic media, having paid due respect to *yours* and *mine*, is to emerge *ours* within an altogether happy family having possessions in common that can only increase in value over time.

## YOURS

First, let us consider the viewpoint of the consumers of systemic media. Who are they, what are their needs, what do they own (or need to own), and what sense can they possibly make of systemic media? We must begin by saying that although this *who* is considered to be consumers of systemic media, that is not who they are. They are in reality a set of stakeholders (they may not be individually aware of their membership of this set) each holding a unique perspective on some system of interest, for example, the IED problem. In part, it is this particular tenure that unites them as a set. The extent to which these idiosyncratic perspectives can sublimate into a shared cultural experience is a key to resolving tensions among stakeholders and creating breakthrough insights for the collective as a whole. This sublimation points to a resolution of individual (and essential) autonomy, not to be understood as surrender, alongside a need for belonging by every stakeholder through which a greater and deeper understanding of the shared system of interest can be obtained.

These stakeholders are problem owners—notwithstanding the lack of ownership commonly found in matters of high complexity (as we will observe in more detail in the next chapter). They are also, at least initially, owners of very different problems, often failing to discern that these personally held views are in some way peculiar facets of some deeper problem common to all. This being the case, it would come as no surprise to discover an attitude among these stakeholders of skepticism if not outright hostility toward the notion that outsiders can help, especially those who talk an arcane language that speaks only to the most abstract affairs. And yet, it is possible, as we hope we have shown on several occasions, to present a systemigram as though it directly captured the affairs that matter to these stakeholders and that represented their various albeit widely differing views on the system of interest. It would indeed be quite a feat for the creators of systemic media to provide a narrative and supporting graphical model that was immediately and wholly embraced by the stakeholders as though it was something they had said, that they had thought, and that they themselves had birthed. Such is the need: problem owners become eager consumers of systemic media because it speaks directly to them and has something original and uniquely valuable to give.

This being so, the creators have to think like the stakeholders, without suffering from the biases, myopia, prejudices, and all other such ails by which the latter are inevitably plagued. The creators must not only emulate stakeholder thought patterns. They must mimic stakeholder vernacular, and they must be immersed in, but not overwhelmed by, stakeholder culture. Only by so doing can the systemic media created by *me* and rightly *mine* be veritably regarded as *theirs*. Therefore, systemic media's verisimilitude must be a compelling factor in the engagement of stakeholders and in their self-organizing congruence.

## MINE

Now let us turn to the viewpoint of the producers of systemic media. Who are these folk? Who is this *me* and what is the *mine* that is not *yours*—and yet is! Perhaps one of the first things we should say about this *me* is that he or she is a systems thinker. So what? Isn't everybody? We are all familiar with the adage "Those who can *do*, those who can't *teach*, and those who can't teach *write books*." Maybe this needs extending with the phrase, "and those who can't write books (very well) write books on systems thinking!" That puts us in our proper place, if true. So even if everybody is a systems thinker, it seems to serve no purpose to argue that this *me* is a professional systems thinker. What was this *me* before being thus categorized? And how might this prior experience serve to explain the value of becoming primarily a systems thinker?

This *me* used to be a fully paid-up member of the world of work. It really does not matter in what professional capacity—science, medicine, technology, management, or finance. In that world, one inevitably encounters all kinds of systems. Some are good, others are bad. Some behave, others are pernicious. Some endure, others are ephemeral. Some adapt and get better, others are rigid and obstructive. Some have agility and resilience—they learn fast and come back fighting after a devastation; others are destined to sleep with the fishes. This *me* knows this about systems. This *me* pointed out these features to colleagues. This *me* was willing to get involved in remedying and enhancing these systems. This *me* encountered frustration and failure to fix problems. This *me* didn't blame anyone for these experiences. This *me* accepted that the answer lies within, but as importantly, it also lies without. So this *me*, realizing that two views were needed and these views needed to be mediated, did something about it. This *me* joined the exterior to help provide that view and to help the exterior become more acceptable to the interior as part of the problem-owning/problem-solving process, which it seldom is. After all, if you leave the world of work it's because you couldn't make it in that world, right? Maybe. But maybe there is another explanation, the one that this *me* can now explain.

What this *me* saw in systems was the possibility that they could be superb—in an ideal world. They could be beautiful; they could exhibit harmony and elegance. They could be magical. They could enable engineering to become *Imagineering*. They could connect people and serve their needs. They could help resolve tensions in individual lives, in social networks, and in national affairs. They could bring freedom and justice. They could enhance life and enable liberty. They could drive the pursuit of happiness. In an ideal world. This *me* knows that for the most part, this was the mission and motivation behind the creation of all the systems in the world of work. For some reason, these systems did not live up to expectations and for many became the butt of jokes and even the objects of hatred. But none of these realities takes one iota of value away from the verity that systems can be and, in an ideal world, are superb. Systems, as an idea, are magnificent. As real objects, they fall short. This *me* knows that by taking an exterior view and mediating this with the interior view, this magnificence can be a beacon of hope and a lodestar of reform.

## OURS

We have now articulated *yours* and *mine* as distinct viewpoint holders, relative to some *system of interest* that emerges from a reflection of the world of work, a world replete in systems and with no shortage of problems, some of them wickedly complex. We have tried to state that the yours viewpoint is an interior view and that the mine viewpoint is an exterior view. Since in reality no one actually leaves the world of work—it's all the same place for all of us—what then is this demarcation? What is the (system) boundary that encloses the interior and by all the laws of mutual exclusion determines an exterior? And it cannot be a boundary that has thinking on one side and doing on the other. Thinking and doing take place simultaneously on both sides of this boundary. What then is it that makes this separation possible and these views distinct?

It is not easy to say what this is with mere words without being held hostage by those very words. Yet something has to be said. And what we suggest is that the interior is a very real world and the exterior is an ideal world. In the former, systems are real objects that have all the characteristics that we mentioned above: good and bad, helpful and hindering, smart and dumb, lasting and expiring. In the latter, systems are merely magnificent. They are admirable, delightful, inspiring, and magical. Do these ideal systems have utility? They do. They bring hope where there is despair, they bring clarity where there is confusion, they bring insights where there is perplexity, and they bring choreography where there is chaos. Great! How?

We chose the term "systemic media" with care. It is not simply a fancier title by which systemigrams are known, although at the outset these devices and their supporting narratives and storyboards are what gives us our first glimpse of systemic media. The term "media" today is taken to be a singular noun to describe the apparatus by which we receive our news (including news of promotions, i.e., advertisements). The media lets us know what is going on in the world. It stands today as a collective noun and therefore can properly be regarded as a singular noun. But the term originally was the plural of "medium," a term that was suggestive of a space or function that went in between two other spaces or functions. Hence, the term "mediation," which derives from the Latin *medius*, meaning "placed in the middle." Our use of the term is very purposeful: we want to leverage the apparatus that today's media utilizes, and we want to mediate between two parties: the realists and the idealists. We assert that the combination of apparatus, for example, film, animation, social IT, cloud computing, mobile devices, and the like, with the delicacies of mediation can in principle deliver a powerful paradigm for addressing complex problems and facilitating their resolution that will produce outcomes vastly superior to current problem-solving approaches.

One final word on this matter of mediation. Systemic media, directed at a peculiarly complex problem, can exist only because its creators—the idealists—birthed it. That makes it *mine*. But there is no way that it can exist without the bitter experiences of the realists. That makes it *yours*. One object (or set of objects), two owners. It is vitally important that the idealists do not regard it as *mine*. Yet they must. It is equally important that the realists regard it as *yours*. Yet they cannot. How we encapsulate this paradox is by saying, "For me to own it is to know it is not mine, and for you to own it is to know it is not yours." It is now *ours*. And what we always intended systemic media should be, a beacon of hope and a lodestar of reform, will naturally and spontaneously be realized by a harmonious community of problem owners, to whom the desire for systemic elegance has been restored, and system mediators, for whom the value and utility of systemic elegance has been confirmed.

## LIGHTS, CAMERA, ACTION: A FILM IN RESILIENCE

Who should bother to create a systemigram? To whom should you show a systemigram? In this section, we explain in detail the systemigram on *What Is Resilience*, how it was created, to whom it was exhibited, and with what outcomes.[1] We highlight insights we gained from its creation, why we were the modelers, and how we did not own the model.

### Preamble

Resilience has begun to dominate the U.S. position on homeland security and specifically the maritime domain. The need for a collective focus and understanding of resilience for the U.S. Department of Homeland Security (DHS) and maritime security was a source of motivation for a research project conducted by the DHS

[1]Parts of this section are from Sauser, B., M. Mansouri, and M. Omer (2011), "Using Systemigrams in Problem Definition: A Case Study in Maritime Resilience for Homeland Security." *Journal of Homeland Security and Emergency Management* **8**(1): 1–19. Reprinted with permission from Berkeley Electronic Press.

Center for Secure and Resilient Maritime Commerce (CSR). We will describe how we used systemic media to help the DHS better articulate, understand, and mobilize resilience. To articulate the roles of yours, mine, and ours, let use explain this in classical terms of stages in filmmaking, that is, *Development*, *Preproduction*, *Production*, *Postproduction*, and *Distribution and Exhibition*.

## Development (Yours, Then Mine)

At the start of this stage, it is about the problem situation being first experienced in its purest essence. This can be based on many presumptions, so every attempt is made not to extrapolate about the nature of the situation. For resilience, we have found a concept or term that seems to mystify many and even cause tension between allies. Thus, the problem of resilience in its unstructured form is, "What Is Resilience?"

Second, we need to formulate a description of the situation within which the problem occurs. Both logic and the culture of the situation are taken into account at this point. With the plethora of literature in a diversity of domains that have described or defined resilience, we have found as many consistencies as differences. But this literature is what gains us the understanding of resilience and how its problems and application are being discussed among a body of scholars and practitioners.

Third, we conceptualize the problem situation in structured text. The structured text identifies the key elements, with attention to systems thinking modeling and analysis requirements. Using this extensive body of literature, we were able to write a document that summarized what we found, what was similar, and what was different—a textual system of the essence of resilience. We made every attempt not to change the original words or thoughts of the authors, but to stay true to the essence of their views on resilience.

Finally, we begin the creation of a systemigram as designed from the structured text to capture and represent the essence of the original conceptual thinking. This was our interpretation of the literature where we made every attempt to consider the multiple perspectives. This initial systemigram is represented in Systemigram 19.1.

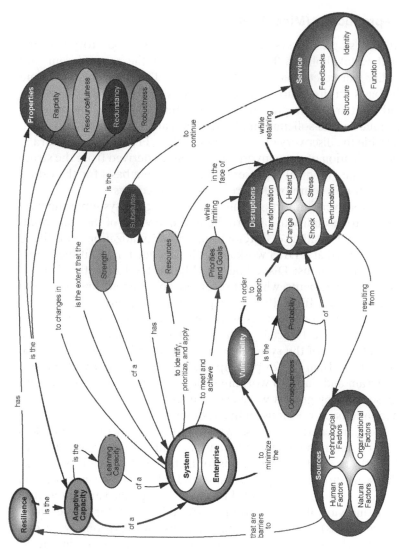

**Systemigram 19.1.** What Is Resilience—"Development Stage"

191

## Preproduction (Mine, Then Yours)

At this stage, the systemigram is dramatized via storyboarding to the stakeholders. This is done so that the systemigram and reality can be compared and contrasted. The differences become the basis for discussion: how do things work, how might they work, and what are the implications? The systemigram diagram provides a venue for the solicitation of individual and group inputs to make possible the discovery of relevant new ideas. From the systemigram, the realization of the convergence of values derived from the structure of the graphical representation can give a basis for the establishment of a common culture across perspectives.

This is a very important stage for verifying the systemigram with respect to its ability to capture the multiple views of the stakeholders. For our resilience systemigram, the dramatization and dialog was executed at a DHS workshop, "Resilience for Maritime Transportation Systems: Dispelling the Myths; Exploring the Truths," with 25 participants who represented government, industry, and academia (see Table 19.1 for represented organizations). At this workshop, the participants were presented with an overview on resilience and a tutorial on the use of systemigrams, as well as a dramatization of the systemigram that resulted from the Development stage. Then the 25 participants were given a copy of the systemigram and put into working groups of four to five participants. These working groups were asked to comment and make recommendations on revisions to the systemigram.

We iterated this process with another group of stakeholders from the maritime homeland security community at a second workshop, "Using Systemic Diagrams for Defining Maritime Resilience." This workshop had 40 participants and followed the same method as previously described, but this time, participants were allowed to work individually or in groups (see Table 19.1 for represented organizations). Again, recommendations and revisions were collected verbally and in written modification to the systemigram.

## Production (Yours and Mine)

At this stage, the identification of feasible and desirable changes is deciphered from the previous stage, understanding that they are

**TABLE 19.1. Workshop Participant Organizations**

ABS Consulting[1]
American Trucking Association[1]
Booz Allen Hamilton[1]
C&H Patriot Security[1]
Council on Foreign Relations[1]
CSX Transportation[1]
DiMatter & Associates, Inc.[1]
Donjon-SMIT[1]
Inland Rivers, Ports & Terminals[1,2]
Massachusetts Institute of Technology[1]
MG Group[1]
Mississippi River Authority[1,2]
New Jersey Office of Homeland Security Preparedness[1,2]
New Orleans Port Authority[1,2]
New York Sandy Hook Pilots[1]
New York State Office of Homeland Security[1,2]
Port Authority of New York and New Jersey[1]
The Port of Long Beach[1]
The Port of Los Angeles[1,2]
Sandler & Travis Trade Advisory Services, Inc.[1]
SSA Marine[1]
Stevens Institute of Technology[1,2]

U.S. Army Corps of Engineers[1,2]
U.S. Coast Guard[1,2]
U.S. Department of Homeland Security[1,2]
  Customs and Border Protection[1,2]
  Domestic Nuclear Detection Office[1]
  Federal Emergency Management Agency[1,2]
  Office of Infrastructure Protection[1]
  Science and Technology Directorate[1]
  Transportation Security Administration[1,2]
U.S. Department of Transportation,[1,2]
  Maritime Administration[1]
University of Minnesota, National Center for Food Protection and Defense[1]
University of North Carolina, Center of Natural Disasters, Coastal Infrastructure and Emergency Management[1]
University of Pennsylvania[1]
University of Puerto Rico-Mayagüez[1,7]
University of Southern California, Center for Risk and Economic Analysis of Terrorism Events[1]
Vickerman and Associates[1]
World Shipping Council[1,2]

[1]Department of Homeland Security Workshop on Improving Port Systems Resilience, Charlottesville, VA, "Using Systemic Diagrams for Defining Maritime Resilience," May 28, 2009.
[2]7th Annual Maritime Homeland Security Summit, Ponte Vedra Beach, FL, "Resilience for Maritime Transportation Systems: Dispelling the Myths; Exploring the Truths," April 30, 2009.

likely to vary. Desirable asks if it is technically an improvement. Feasible asks if it fits the culture. As a result of the preproduction stage, stakeholder recommendations and revisions were compiled in order to create a new version of the systemigram and the structured text.

## Postproduction (Ours)

At this stage, every individual or collective input that is deemed desirable or feasible is incorporated into a revised systemigram. Only contributions that answer "no" to one of the two questions of desirable and feasible are dismissed. The value of this work was to create a systemigram that could convey greater meaning and relevance to our understanding of what is resilience. At this stage, it is important to achieve a systemigram whereby (i) the people concerned, that is, stakeholders, feel that the problem has been solved/defined; (ii) the problem situation has been improved; or (iii) insights have been gained. Thus, success was determined by the DHS customer and their key constituency when they determined we had reached an acceptable level of (ii) and (iii) and thus no further iterations were necessary.

## Distribution and Exhibition (Ours and Theirs)

So now we have our film, that is, systemigram and supporting scenes. The value of this work was to create a systemigram that could convey greater meaning and relevance to our understanding of what is resilience. Therefore, the systemigram now belongs to the stakeholder to use this information in developing better methods, processes, and tools that can systemically address the complex nature of resilience. Likewise, with distribution and exhibition, the systemigram now belongs to everyone—theirs.

For the systemigram depicted in the Final Scene (Systemigram 19.13), we decompose it into scenes that represent constructs in the literature entitled *Resilience* (Systemigram 19.2), *Adaptive Capacity* (Systemigram 19.3), *Systems of Resilience* (Systemigram 19.4), *Vulnerability* (Systemigram 19.5), *Disruptions* (Systemigram 19.6), *Resilience Properties* (Systemigram 19.7), *Rapidity* (Systemigram 19.8), *Resourcefulness* (Systemigram 19.9), *Redundancy*

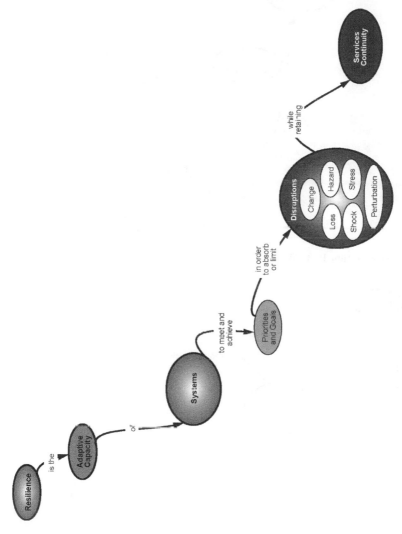

**Systemigram 19.2.** What Is Resilience—Scene 1: "Resilience"

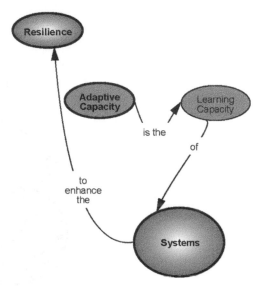

**Systemigram 19.3.** What Is Resilience—Scene 2: "Adaptive Capacity"

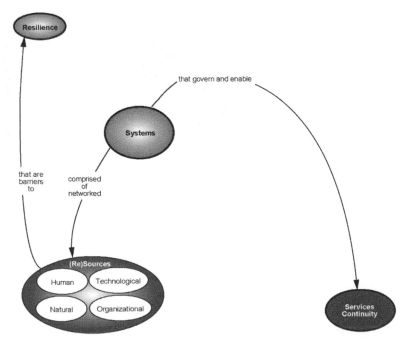

**Systemigram 19.4.** What Is Resilience—Scene 3: "Systems of Resilience"

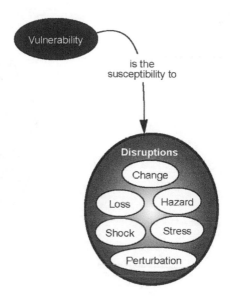

**Systemigram 19.5.** What Is Resilience—Scene 4: "Vulnerability"

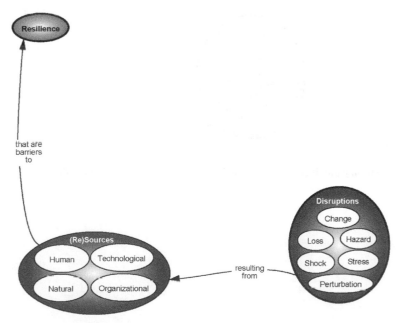

**Systemigram 19.6.** What Is Resilience—Scene 5: "Disruptions"

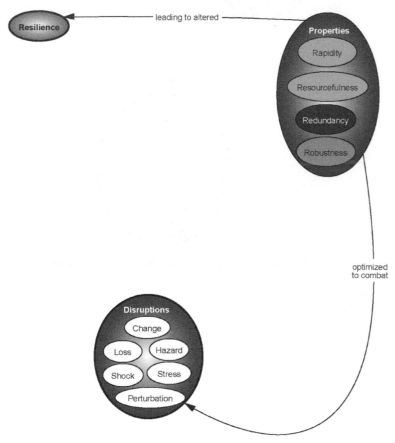

**Systemigram 19.7.** What Is Resilience—Scene 6: "Resilience Properties"

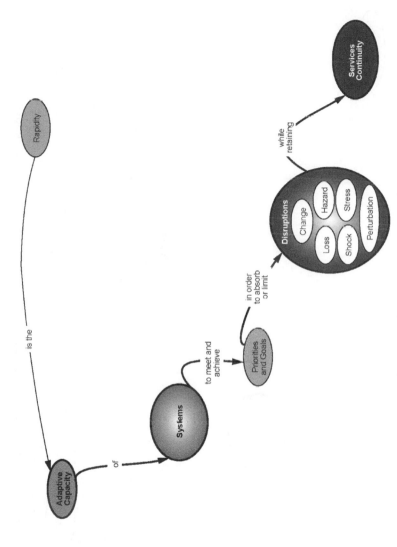

**Systemigram 19.8.** What Is Resilience—Scene 7: "Resilience Properties—Rapidity"

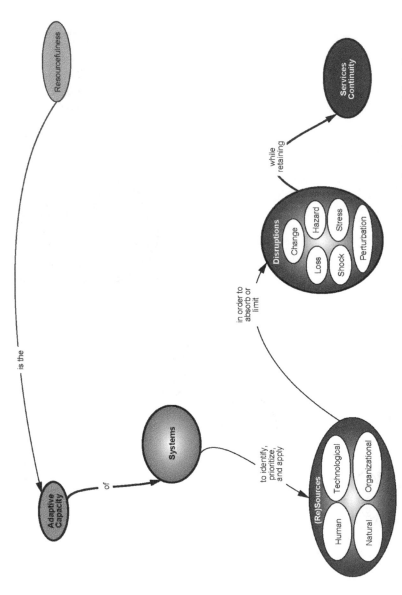

**Systemigram 19.9.** What Is Resilience—Scene 8: "Resilience Properties—Resourcefulness"

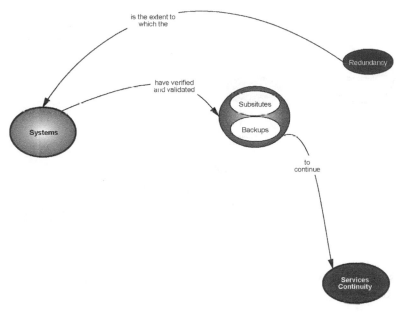

**Systemigram 19.10.** What Is Resilience—Scene 9: "Resilience Properties—Redundancy"

(Systemigram 19.10), *Robustness* (Systemigram 19.11), and *Service Continuity* (Systemigram 19.12). Each scene is then composed of at least two nodes, and any two scenes can share one or more nodes. As an example, Scene 2 on *Adaptive Capacity* (Systemigram 19.3) would read, "Adaptive Capacity *is the Learning Capacity of Systems to enhance the Resilience*," and Scene 8 (Systemigram 19.9) on *Resourcefulness* would read, "Resourcefulness *is the Adaptive Capacity of Systems to identify, prioritize, and apply (Re)Sources in order to absorb or limit Disruptions while retaining Service Continuity*."

Scene 1 (Systemigram 19.2) is our Mainstay, which is the core purpose of the SoI that runs from the top left to the bottom right. It starts with the SoI and ends with the SoI's goal.

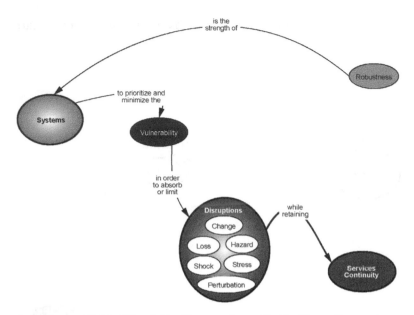

**Systemigram 19.11.** What Is Resilience—Scene 10: "Resilience Properties—Robustness"

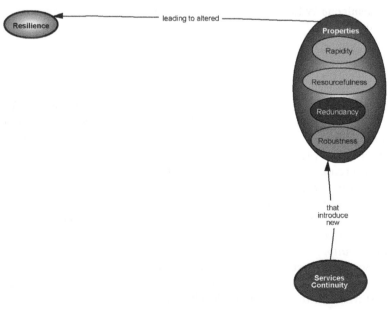

**Systemigram 19.12.** What Is Resilience—Scene 11: "Service Continuity"

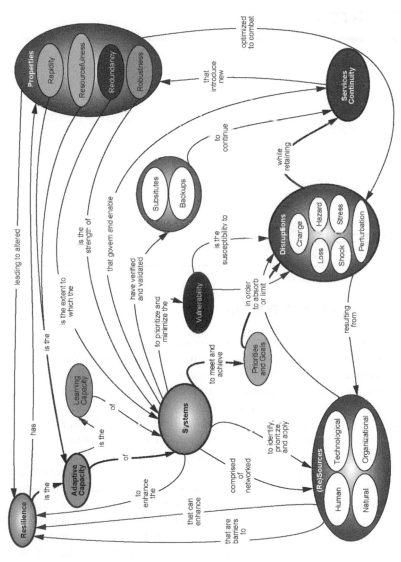

**Systemigram 19.13.** What Is Resilience—Final Scene: "What Is Resilience"

# CHAPTER 20

# WHERE . . . ?

## THERE'S A PLACE FOR US

Ireland is a beautiful country. Notwithstanding its current economic difficulties, it remains a huge tourist attraction to many from all around the globe. And the Irish are so welcoming and helpful. If you are motoring around the quiet country lanes admiring the scenery and unafraid that you might get lost, there's always the certainty of being able to inquire of a local with total assurance of receiving unerring advice. Take the case of one Englishman, Giles Beautement, on a much needed escape from his busy London city life and pressures and gliding along with gentle ease and unfettered tranquility. As it draws near to dark, Giles recognizes his need to find a suitable way to Balbriggan, and happens by a friendly native. Pulling his magnificent Mercedes S320 over by the side of the road, he inquires of his perfunctory guide, Mr. O'Reilly, who courteously lays down his shears to retire from his labors in cutting the hedge.

*Systemic Thinking: Building Maps for Worlds of Systems*, First Edition.
John Boardman and Brian Sauser.
© 2013 John Wiley & Sons, Inc., Published 2013 by John Wiley & Sons, Inc.

"Could you tell me the way to Balbriggan, please?" asks Giles with typical English etiquette. Mr. O'Reilly wipes his brow. "Certainly, sir. If you take the first road to the left? No still that wouldn't do now! Drive on for about four miles then turn left at the crossroads. No that wouldn't do either for sure." Mr. O'Reilly scratches his head thoughtfully. "You know, sir, if I was going to Balbriggan, I wouldn't start from here at all."

An old joke. Yet one that addresses a great many contemporary situations in the matter of solving the kinds of problems about which we have been writing. We are where we are, that's for sure. And we always know where we need to be. But do we know where we are? And are we willing to travel to where we want to be from where we are? In this chapter, we consider systemic media to be a journey, one that begins at a place where problems are not exactly how they first seem and which are themselves beset by ancillary niggles that can prove even more troubling than the problems we are trying to address. This journey does not conclude, as traditional problem-solving approaches do, with a solution. It continues from a phase of collective enlightenment among the community of problem owners through the program of shared executive action this community determines to a destination it determines in the light of a fresh awareness of each other's needs.

## THE PROBLEMS WITH PROBLEMS

The primary purpose of all living creatures is survival. After that comes food. And for the brave (or foolhardy) few, then come thoughts of empire. But survival comes first. It is this context that gives rise to our fundamental view of the world: that out there are problems, that problems are to be solved, and that once solved, we then move on to the next problem and its solution. The process is interminable, even for the emperors. It is therefore only natural that when we see a problem, it is there to be solved. Not debated nor contemplated but solved. All of this is posited on the twin notions that the problem is known (or knowable) and that the solution deals with it (by eradication, expiration, elimination, or plain exhaustion—in other words, the problem gives up or goes away simply as effort to find a solution is expended).

Survival is both a pressing need and a measure of success. If life survives, whatever problems that had appeared have been solved. If not, then the problem, so far as the dead are concerned, no longer exists. Either life goes on and problems come and go, or life comes to an end and with it the end of all problems—and all solutions. Life in this view is merely a series of problems and solutions lightly decorated with meals and, for some, a set of clothes that may or may not be invisible. This is an inexorable process and the thought pattern it conveys quite unshakable. And yet life is fundamentally about shaking up thought patterns, for if not, the survival mechanism itself must die.

The particular living creatures described in this book are enterprises—firms, small businesses, large corporations, alliances, partnerships, and value webs. These too must survive, find food, and build empires. And so they are not excluded from this "problem–solution–next problem" paradigm. But there is something peculiar about this kind of creature that forces a halt to this pattern and a fresh challenge to the creature's survival mechanism, and it is this: there are problems with problems, what we might term secondary problems piggybacking on what matters to us most, namely, the primary problems. These secondary problems matter. To ignore them and focus on the primary problem is to risk nugatory or meaningless effort. It can even make matters worse. Imagine that sincere effort adding to the burdens that it labors so diligently to remove. It is as though the survival mechanism speeds the death of the creature as opposed to preserving its life.

We see three types of secondary problem: the lack of problem ownership, the lust for silver bullets, and the lack of solution ownership. We will illustrate these three types with reference to the IED problem described in the "How" chapter.

Whose problem? It is relatively simple to create a bubble in a diagram with a label in it that says "IED Problem." The bubble then belongs to the diagram and is connected by very evident links that also bear labels. But whose problem is the IED problem? Certainly the U.S. Department of Defense (DoD) comes to mind. It is headquartered in the Pentagon in Washington, DC, where relatively few explosions occur, thought they did so dramatically on 9/11. It is also a problem—of a different kind—for the noncommissioned officers (NCOs) and corporals targeted by the insurgent enemy. At a higher level, it is a problem for the Commander in Chief, the elected representative of the people of the United States, and for his generals. Curiously enough, it is also a problem for the insurgent enemy who need to know if their strategy and tactics are working. Who owns this problem? All of the above. These groups share the problem but not the same one. Finding a single problem owner is a problem in itself. Perhaps each owns different problems and the idea that there is one single IED problem is a myth. Perhaps each group has a facet of the IED problem? But if so, a systems thinker is bound to ask, what makes these facets and what makes them come together as one or be derived in part from the whole? That is one aspect of the lack of problem ownership—the multiplicity of problem owners and their lack of cohesiveness. But there is a second aspect, which is that of a problem owner shirking his or her responsibility to own the problem because it is merely a facet of a greater problem and the owner, for whatever reason, feels incapable of dealing with that facet. Perhaps because to do so would make matters worse for others, or because others need to act in order to make their facet less complex. Some people do not realize that the problem that falls to them is truly theirs. Perhaps because in their view it was someone else who created it and that person must be the one to solve it. Or perhaps because they refuse to accept what others say is a problem, being reluctant to act at the behest of those they see as "accusers" or "troublemakers." The secondary problem of problem ownership is endemic in all enterprise creatures.

## Where Wolf?

A second type of secondary problem is the lust for silver bullets. When a problem has thus far proved intractable, it is not uncommon to place maximum effort into a breakthrough solution, be this in the form of new technology or technique, for example, algorithm. Just as the werewolf in folklore is slain only by this unique form of ballistic, so problems of the kind we have been describing warrant a unique form of solution. The metaphor is apt. The bullet will surely slay our monster, but only if we aim correctly. At what do we aim? What is our target? What exactly is the problem? Are there many monsters? Does the multiplicity constitute a single target? And is this single target synthesizable from the multiplicity? This emphasis on "solution determination," the fabrication of a silver bullet, belies the real need, which is to recognize the "problem of problem definition." While this in no way detracts from solution determination, it does, at least respectfully, defer that effort that then gives time to the contemplation of the nature of breakthrough thinking, that then leads to original technique or innovative technology. How does this play out for the IED problem?

Curiously, the IED problem would not occur if there were no insurgents. Insurgents are only possible given a spirit of insurrection. Insurrection arises because of disaffection with government. Where government lies in the hands of nonnative peoples, for example, an occupying force or nation-creating capability, disaffection is more than likely. In effect, soldiers who come in peace, albeit armed, immediately become targets. The broader target is represented by those who send them, the military commanders and peoples in foreign and despised lands. Do these take a bullet to themselves? And if they do so, can they survive it while slaying the IED monster? We assert that systemic media offers a theater for this narrative and more importantly a unique opportunity to produce a conclusive and happy ending.

## I Solution

A final type of secondary problem is that of the lack of solution ownership. The temptation with a multiplicity of problem facets that has no synthesizable single solution is to propose a partial solution based on some primacy of problem facet. This is well illustrated by the account of four blindfolded men placed in a room containing a single object and located at different sections of that object in order to get a sense of what it might be, with each in turn reporting it to be a spear, or alternatively a snake, an oscillating fan, or a tree trunk. None can agree that the object is in fact an elephant because none expends effort in removing his or her blindfold, all having failed to make sense of what each sincerely reports. We do not castigate this situation. It is all very natural and commonplace. It is, however, not acceptable. Accordingly, it must come as no surprise that where there preexisted a general lack of ownership of the problem, resulting in conflicting secondary problems, there can be scant support for what all will see as a myopic and skewed solution, all, that is, save the one who promotes it. This paucity of agreement and subsequent nonexistent (or tepid) support for the implementation of a "glocal"[1] solution would be bad enough. But it can get worse. Malicious compliance is a term we use to demonstrate how some might disingenuously pledge support to a glocal solution while clandestinely conspiring against it and conceivably seeking to bring down, in time, those who "forced it upon" the congregation. The enemy within does not help to overcome the one without, who may even be smart enough to know how to exploit this secondary problem and make it a much larger one than the primary problem created by the external enemy in the first place.

## Way to Go

The realization that a primary problem gives rise to secondary problems of the kind we have described tells us where we are, and therefore from where we must begin our journey to our chosen destination. Our initial location is defined not solely by the primary problem but also by the culture in which this problem is enmeshed and which in a vital sense gives it its energy. The totality of the problem and its culture must be understood if we are to properly identify our starting point. But in describing this totality, we provide ourselves with some excellent signposts that then inform us of directions toward our destination—the resolution of the primary problem.

The first of these signposts, commensurate with the need to embrace the problem's totality, is to *stand back*. Relative to the IED problem, standing back enables us to see, for example, that

---

[1]We are indebted to the movie *Up in the Air* for this term. We use it in the sense that a local solution—to a problem facet—is to be confirmed as one that satisfies a global problem.

each letter of the acronym "IED" is lethal, with perhaps the letter "I" being the most lethal if not the most immediate concern. Armor and materiel is certainly necessary to shield against the explosion, the "E." But intelligence is necessary to combat the devices, the "D," in terms of acquisition of materiel by the enemy, distribution channels, both financial and logistical, for the device's fabrication and deployment. Who knows what is needed to respond to the letter "I"? Certainly, an improvisation (or agility) on our part coupled with foresight to anticipate the next level of threat, wisdom to know why it persists, and will, strengthened by knowing and holding on to our own values, to combat that threat.

A second signpost is to *admit complexity*. What this means is to have a full and proper acknowledgment that what you are faced with is not something that is complicated but is complex. The two are different. The former can be resolved by an unfolding so that the intricacies of what is first seen are laid bare as something more elementary. The latter cannot be unraveled because everything is connected to everything else, giving influences a viral nature, which is to say the effects of an influence ripple through the complex and begin to change the initial influence in unpredictable, unforeseeable ways. A complex can neither be unfolded nor unraveled, and any such attempts to do so are thwarted from the outset.

But we are not left in this condition, paralyzed as it were by an immutable impotence to affect the complex. A third signpost avails: the faith that behind complexity lies a subtle simplicity, that to understand overt complexity is to *seek simplicity*. Imagine the IED problem as a social network; the systemigram foreshadows this. The nodes in the network are a mixture of human agencies (e.g., insurgents), artifacts (e.g., multi-role armored vehicles, or MRAVs), and attitudes (e.g., belief systems). The links are the logical explanations for these nodes to be interconnected according to some world view. As a social network, regardless of how complex it becomes, it will inevitably assume an underlying signature. The two types of signature that are most well known are *aristocratic* and *egalitarian*. The former is explained by the forces of growth and preferential attachment and leads to super hubs that are critical to the network's virility. The latter is characterized by huge numbers of tightly coupled clusters and a paltry scattering of weak links connecting distant clusters. In both cases, an immensely large and interwoven network is reduced to a small world, whereby any node is reachable from any other by a handful of direct links. What this shows us is that behind the evident complexity of large networks lies the hidden simplicity embodied in either a few strong hubs or a few weak links, either one acting as super connectors. This signpost gives the clues as to where in a complex problem lie the crucial elements on which to focus management attention.

A fourth signpost is fully consistent with the first, to stand back, and the second, to admit complexity. In a curious way, it also reinforces the third, which is to seek simplicity (in the face of bewildering complexity). It is this: to *honor perspectives*. Inevitably, as one stands back from the kinds of problems we have been describing, an increased number of viewpoints or perspectives is being admitted. More stakeholders have their say, and of course each is valid, at least so far as the holder of that viewpoint is concerned. Everyone is correct, but logically, they cannot all be correct, which is a clear admission of complexity. What blade can cut this Gordian knot? And how do we mere mortals assume the mantle of Alexander's greatness?

If the perspectives, conflicting, competing, or otherwise, are considered valid, then maybe that validity must be assumed and shared. This is to honor perspectives. It permits an inquiry into why these particular perspectives are held and how firmly or tenuously they are clung to by their holders. Such an inquiry, conducted humbly and respectfully, is one directed at capturing the culture of the problem. It is an inquiry that seeks to understand the validity of the perspective holder and in particular that holder's *validity to belong* to the entire community of viewpoint holders. The perspective holder asserts his autonomy, rightly so, and one clear manifestation of that autonomy is the holding of a specific perspective, held vigorously and unchangeably perhaps. But the *simultaneous validity* of all perspectives requires of any single autonomous viewpoint holder reason to belong. Leveraging that reasoning provides opportunity to develop shared perspectives and therewith a subtle simplicity that lies behind the evident complexity.

There is one final signpost. It is to *appreciate emergence*. Living creatures in the natural world are said to evolve, to adapt, and to learn. These are processes that social creatures, such as the enterprises we explore with our systemic media, do well to emulate. Corporations that understand and can work with evolution, by adaptation and learning, stand a much better chance of occupying the landscape for longer and coping with its upheavals more resiliently, coming back from adversity with greater powers of endurance, endeavor, and enterprise. Emergence is what happens when things come together, when systems combine, when corporations partner, when product and service teams form. Some of this emergence is deliberately intended, some is unintended, and can be serendipitous or unfavorable. Emergence is an inescapable reality in complex systems and problems. Knowing about it will help us get and keep our bearings; it will comfort us when the place we are trying to get to is not best reached from where we are. If all of these five signposts are respected, then the problems with problems will come as less of a surprise, and dealing with the primary problems will go largely unthwarted by the occurrence of the secondary problems.

## A SCENARIO

And so now we know our origin—a place where problems are more complex than they first appear and which are beset by irritating niggles that bite us if ignored, making the solution of these perceived problems more troublesome than if we had left them well alone in the first place! And we know our destination, the shared satisfaction of the community of problem owners with the efficacy of their collective executive action, a program enlightened by the use of systemic media.

The apparatus of systemic media comprises for any given System of Interest (SoI): the SoI description, the systemigram, the System-iShow, and the bringing together of these artifacts with the community of problem owners for a facilitated vigorous and open debate directed toward shared learning. It is quite appropriate, therefore, to regard systemic media a vehicle for the journey. We are bound to ask, however, "Is there yet more value that systemic media can add other than to shed light on the confusion among problem owners and to draw admiration for the elegance of the systemigram and its accompanying presentation?" We believe there is.

We have found that systemigrams invite problem owners seeking to know how best to take action to consider a variety of "what if" scenarios and by so doing to inspire them in the definition of the program of executive action they plan to follow. Let us look once more at the IED systemigram and use it to pose just one scenario.

First, notice that at the very center of the diagram lies the node "explosive kills." This should come as no surprise. If this node could be removed, the problem would go away, surely? This suggests a number of possibilities. Clearly, if there were no insurgent enemy, there would be no explosive kills. Yes, but there are, and their removal poses a complex problem if approached by a mirror image "kill" strategy. Next, we might consider the disruption, dismantling, and destruction of the channels of labor by which the kills occur— the procurement and deployment strategies by which the IEDs are assembled and located for enemy action. This approach is no simpler than that required for removal of the insurgents. No one said it would be easy! The proximity of the "explosive kills" node to the "DoD" and the "device mentality" nodes is significant. It is a proximity that reinforces the alleged conventional mindset of the DoD and subsequent emphasis on "traditional weapons of war." Focusing on this proximity can cause us to overlook an important systemic feature, namely, that the "explosive kills" node has no direct connections to the upper half of the systemigram; instead, it has one input—from "intense improvisation"—and two outputs— to "USA & Allies resolve" and to IED learning culture. Let us study these links and their significance.

What the IED systemigram is telling us is that these "explosive kills" are *merely* a lever, one that is pulled by the insurgents' extraordinary agility to stay ahead of the curve regardless of what the United States and its allies do in response to previous scenarios. And the motive for this leverage, the reason why the insurgents deploy this tactic, is twofold: to weaken the resolve of the enemy they fight and to learn better and faster what next they need to do to keep this mission going, toward the insurgents' own satisfactory conclusion. If true, this is remarkable. Killing American soldiers and those of other nations is *merely* a by-product. Making the nations who send them no longer want to do so is the goal, and accelerating toward that goal, by extracting maximum learning from the tactics, is a subgoal. Now that we recognize this, the attack must be on these links. By all means, defense against the node "explosive kills" must remain, but new, more relevant questions must be posed, such as, "What must the United States and its allies do to impede the effect of weakening their resolve?", "How can the IED learning culture (which reinforces their extraordinary agility or intense improvisation skills) of the insurgents be impaired or interfered with?", and, lastly, "How can we cut the link or corrupt it so that the lever loses its effectiveness?" These are not easy questions, but they may be the better ones to address. If so, then systemic media must take some credit for telling us where we are and where we might head if we want to get to our own goals.

## THE PROBLEMS WITH PROBLEMS: DHS SMALL VESSEL SECURITY STRATEGY

Strategies of the U.S. Department of Homeland Security (DHS) have articulated that the U.S. homeland security solutions will be found in the national enterprise only via a collective and shared responsibility that stretches from national to local to community involvement—an enterprise.[2] A fundamental challenge is that while the problems DHS faces may be identifiable, some have argued that there is a problem in defining the problem. Among various emerging homeland security threats, the challenges in the maritime port enterprise have become a growing DHS concern due to their significant impact on border security. The potential use of small vessels, such as commercial fishing vessels or recreational boats, in terrorist-related activities has become of increased

[2]Parts of this section have been reprinted from Sauser, B., Q. Li, J. Ramirez-Marquez (2011), "Systemigram Modeling of the Small Vessel Security Strategy for Developing Enterprise Resilience." *Marine Technology Society Journal* **45**(3): 88–102. Reprinted with permission from the Marine Technology Society Journal.

concern and one for which DHS is actively seeking solutions. Human casualties, financial losses, environmental damages, as well as the political impact that have been caused by some small vessel attacks, for example, *USS Cole* in 2000, *M/V Limburg* in 2002, and *MV Faina* in 2008, have driven plans and strategies to fill this security gap. Therefore, the DHS published the "Small Vessel Security Strategy" (SVSS) in April 2008, in which four typical small vessel threat scenarios are defined: domestic use of waterborne improvised explosive devices (WBIEDs), conveyance for smuggling weapons (including weapons of mass destruction [WMD]) into the United States, conveyance for smuggling terrorists into the United States, and waterborne platform for conducting a stand-off attack (e.g., man-portable air-defense system (MANPADS) attacks). The goal of the SVSS is (DHS 2011):

[to] enhance maritime security and safety based on a coherent framework with a layered, innovative approach; develop and leverage a strong partnership with the small vessel community and public and private sectors in order to enhance maritime domain awareness; leverage technology to enhance the ability to detect, infer intent, and when necessary, interdict small vessels that pose a maritime security threat; and enhance cooperation among international, Federal, state, local, and Tribal partners and the private sector (e.g., marinas, shipyards, small vessel and facility operators), and, in coordination with the Department of State and other relevant federal departments and agencies, international partners.

The development and publication of the SVSS has yielded considerable review and discussion of small vessel security and debate on where are the problems. As an enterprise problem, we looked at the problem of identifying the problem with the SVSS. This kind of information about an enterprise can be essential for effective governance, security management, or resilience in homeland security. As an enterprise model, systemigrams can provide stakeholders with comprehensive knowledge about the architectural structure of the extended network of activities in their environment. This comprehension can equip system analysts with relevant information with which to understand systemic issues, such as bottlenecks in organizational processes and communications. Respective scenes from the systemigram model depicting the Strategic Foundations, Strategic Environment, and Strategic Vision are dramatized to better understand and identify the significant elements within the small vessel security enterprise as articulated by the DHS SVSS. Collectively, these scenes represent a systemic description of the SVSS.

Our systemigram model of the DHS SVSS as defined by the respective SVSS document yielded 12 scenes, depicted in Systemigrams 20.1–20.12. Each scene is a representation of a defining topic within the document. That is, Scenes 1–4 define the Foundations of the SVSS: Purpose (Mainstay), Scope, Relationships to Other Strategies and Plans, and Methodology; Scenes 5–10 characterize the SVSS Environment: Importance to the Maritime Domain, Maritime Governance, Small Vessel Community, and Small Vessel Risk; and Scenes 11–12 articulate the SVSS Vision: Major Goals. Collectively, these scenes represent a systemic description of the SVSS and are collectively the systemigram model in Scene 1 (Systemigram 20.13). We will now explain the essence of each scene.

## Foundations of the SVSS

Every strategy begins with a purpose that articulates the existence of that purpose, for whom it exists, and the outcome of that existence. As the Mainstay, Scene 1: Purpose (Systemigram 20.1), starts with the system of interest (*DHS Small Vessel Security Strategy*) and ends with the outcome of the system of interest (*Adequate Security, Fundamental Freedoms, and Economic Stability*). To read the Mainstay from beginning to end is to read a statement of purpose of the system of interest. Scene 2 (Systemigram 20.2) represents the Scope of the SVSS. We can observe from Scene 2 that while the scope does not include the goal of the SVSS, it does include the other nodes from the Purpose, with the addition of nodes representing the stakeholders in a *Unified Effort*. Fundamentally, a scope should contain the features and functions that characterize a product, service, or result. While one may argue that the scope represented in Scene 2 could be expanded, it is fundamentally the essence of the SVSS scope in that it identifies the

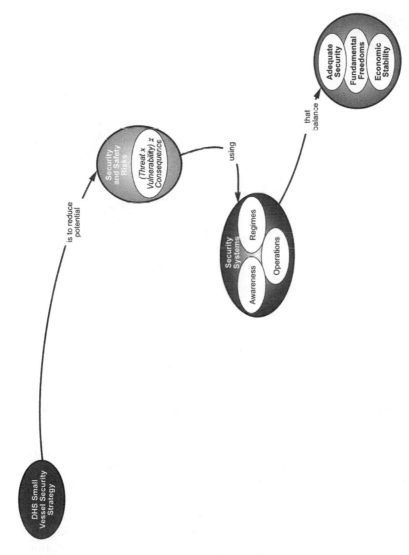

**Systemigram 20.1.** DHS Small Vessel Security Strategy—Scene 1: "Purpose"

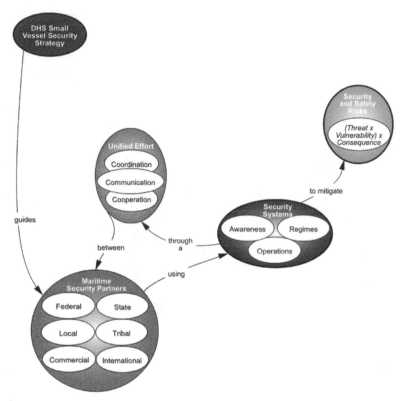

**Systemigram 20.2.** DHS Small Vessel Security Strategy—Scene 2: "Scope"

features of *Security Systems*, *Maritime Security Partners*, *Security and Safety Risks* and the functions articulated by the linking verb phrases, as well as the node *Unified Effort*. Scene 3: Relationships to Other Strategies and Plans (Systemigram 20.3) represents that the SVSS has an alignment with current *DHS Legislation and Strategies*, and the SVSS does not work independent of this guidance. Scene 4: Methodology (Systemigram 20.4) is an articulation of the analysis of the principles that represent the SVSS. For this case, it is the analysis of the *Risk Scenarios* of small vessel security via the *Security and Safety Risks (Threat × Vulnerability × Consequence)*.

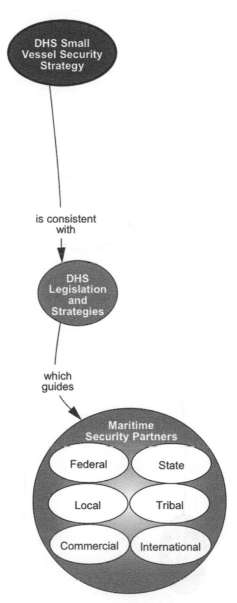

**Systemigram 20.3.** DHS Small Vessel Security Strategy—Scene 3: "Relationship to Other Strategies and Plans"

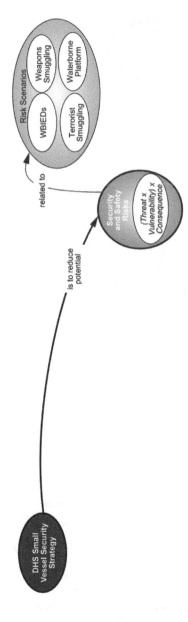

**Systemigram 20.4.** DHS Small Vessel Security Strategy—Scene 4: "Methodology"

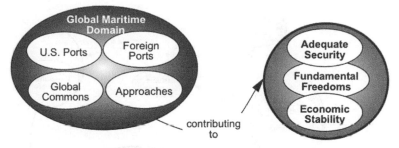

**Systemigram 20.5.** DHS Small Vessel Security Strategy—Scene 5: "Importance of the Maritime Domain"

## SVSS Environment

An environment refers to the surroundings of a system of interest, and for the SVSS, this is the stakeholders and the identified risks and management of those risks as they relate to the SVSS. Scene 5: Importance to Maritime Domain (Systemigram 20.5) represents the direct relationship the *Global Maritime Domain* has with the goal of the SVSS. While other scenes will explain the relationship that the *Global Maritime Domain* has with other perspectives in the systemigram, it is key that this constituency is directly contributing to the goal of *Adequate Security, Fundamental Freedoms, and Economic Stability*. Governance is to steer an organization or set of constituencies based on established or customary guidelines for actualizing a desired status (Mansouri et al. 2010). Maritime Governance, as depicted in Scene 6 (Systemigram 20.6), represents the stakeholders and their actualization of *Risk Mitigation Alternatives* and the *Safety and Security Risks*. A key observation of this scene is that the *Small Vessel Community* is absent. This observation will be evident in two later scenes on the *Small Vessel Community* and Major Goal A. In addition, we will discuss some implications to this observation in the Conclusions and Future Directions section. Scene 7: Small Vessel Community (Systemigram 20.7) depicts the *Small Vessel Community*, but more noticeably, as articulated in the SVSS, the disconnect of the *Small Vessel Community* with the *Maritime Governance* and *Maritime Security Partners*. While this is a reality in practice (see Podlich 2009), the SVSS and other DHS documents, for example, DHS (2010, 2011), explain that this is not

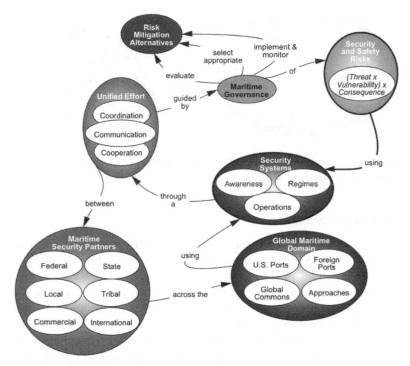

**Systemigram 20.6.** DHS Small Vessel Security Strategy—Scene 6: "Maritime Governance"

a long-term objective of DHS or the SVSS. Scene 8: Small Vessel Risk (Systemigram 20.8) shows the small vessel risks in the *Risk Scenarios*, and their relationship to the *Small Vessel Community* and the other risk-related nodes.

### SVSS Vision

Fundamental to any strategy is a vision that is defined by its goals, that is, "What you want to accomplish?" The SVSS has defined four major goals, and the ability to model these goals as an integrated part of the entire strategy via a systemigram allows us the ability to see the interrelationship of the strategy fundamentals and environment to the goals. Scene 9: Major Goal A (Systemigram 20.9)

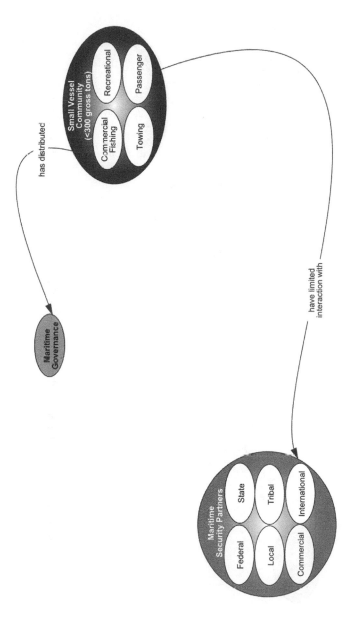

**Systemigram 20.7.** DHS Small Vessel Security Strategy—Scene 7: "Small Vessel Community"

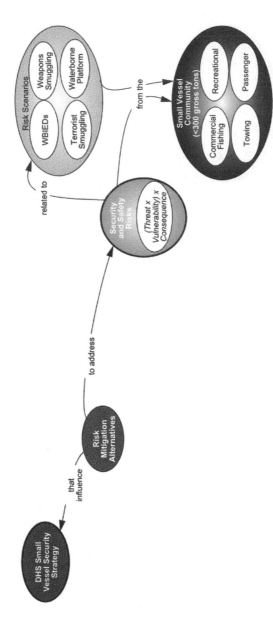

**Systemigram 20.8.** DHS Small Vessel Security Strategy—Scene 8: "Small Vessel Risk"

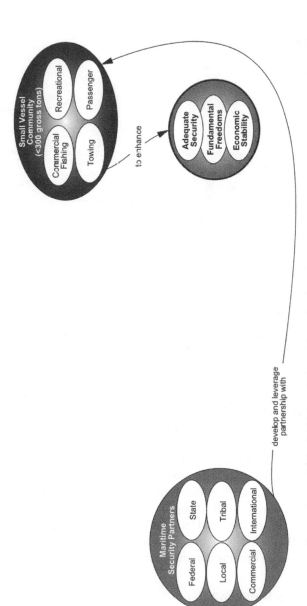

**Systemigram 20.9.** DHS Small Vessel Security Strategy—Scene 9: "Major Goal A"

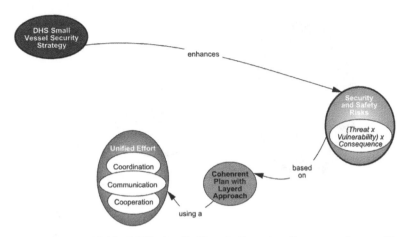

**Systemigram 20.10.** DHS Small Vessel Security Strategy—Scene 10: "Major Goal B"

is related to the DHS commitment to engaging the community as part of the enterprise (DHS 2010) to maritime domain awareness. Scene 10: Major Goal B (Systemigram 20.10) is about defining a plan by which the strategy can be executed. Scene 11: Major Goal C (Systemigram 20.11) is for the development of technology and innovative solutions to address small vessel security. Scene 12: Major Goal D (Systemigram 20.12) has similarity to Major Goal A but is about creating a unified effort with the governing bodies that will execute the plan based on the strategy. Key to each of these goals and the SVSS is a plan that will begin to lay guidance in actualizing the strategy. Therefore, in January 2011, DHS released the first "Small Vessel Security Implementation Plan: Report to the Public" (for purposes of security, a more detailed plan was not released to the public) (DHS 2011). In the next section, we will discuss further what we learned from the systemigram model and how it relates to the implementation plan.

## Where Is the Problem?

What we learned in the analysis of the small vessel security strategy is that the problem has problems—who owns the problem, where

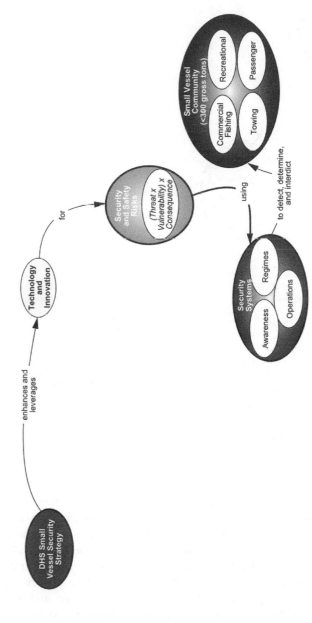

**Systemigram 20.11.** DHS Small Vessel Security Strategy—Scene 11: "Major Goal C"

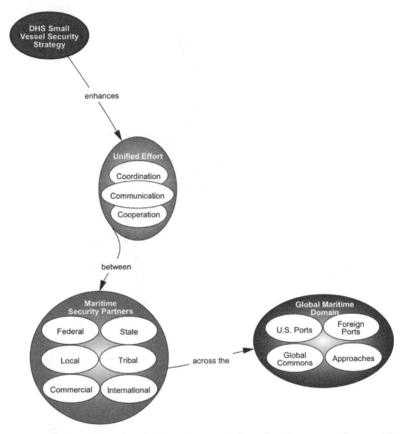

**Systemigram 20.12.** DHS Small Vessel Security Strategy—Scene 12: "Major Goal D"

is the solution, and who owns the solution? DHS released the *Small Vessel Security Implementation Plan: Report to the Public* (the *Plan*) in January 2011 (DHS 2011). Generally, the *Plan* takes the DHS SVSS's four major goals and their associated objectives, and aims at providing more details on identifying related activities and how programs may be developed and coordinated to achieve these goals. If we interpret the SVSS systemigram as an architecture of the SVSS, then we can use this as a framework to assess how

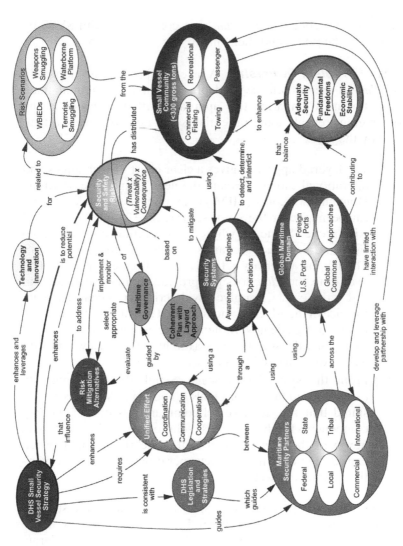

**Systemigram 20.13.** DHS Small Vessel Security Strategy—"The Strategy"

consistent and coherent the *Plan* is in conjunction with SVSS by comparing the nodes and their underlying relationships.

**Lack of Problem Ownership.** A layered approach is adopted in the *Plan* to create defense in depth against the potential small vessel threats. Comparing this approach with the scenes in the systemigram reveals some potential issues not effectively articulated by the *Plan*:

(a) The layered approach gives a clear chart of techniques and operational capabilities and the need to increase Maritime Domain Awareness (MDA) among the stakeholders. These can be mapped in the Scope scene (Scene 2) and Small Vessel Risk scene (Scene 8), where the definition of small vessel risks and four risk scenarios are also given by the SVSS. Though the adversary actions by small vessels are identified over time, the *Plan* does not provide an analytic method to assess these risks, nor give any description of characteristics for each of the adversary actions. That is, as an overall method, it does not guide government agencies, which are supposed to manage the specific risks, as to how the risks would vary when adversary action changes over time.

(b) From the systemigram scene, Small Vessel Community (Scene 7) and the Major Goals A and D in SVSS, it is clearly stated that the small vessel community is one of the key components for the enterprise solution of small vessel threat; however, the *Plan* does not integrate this element when considering their interagency operations that support maritime homeland security.

Thus, while the problem is SVSS, it is not clear in the strategy or the *Plan* who owns the problem. Even if we argue it is an enterprise problem owned by all stakeholders, the systemigram reveals that the collective relationship of those stakeholders is not reticent.

**Lust for the Silver Bullet.** In the section of the *Plan* on "Goals, Objectives, Actions and Program Highlights," the *Plan* states several objectives and suggested example activities for Goal 1, which is associated with systemigram scene Major Goal A (Scene 9) based on the SVSS. Although the *Plan* gives detailed suggestions on how DHS and its components collaborate with the broad maritime community, current programs still have limits to the reporting ability of the small vessel community and the general public. They are also able to offer emergent medical treatment for victims, provide evacuation transportation vehicles, assist with crowd control, and offer instant communication in the event of a small vessel attack. In related research, we have identified these participants and their actions as Zeroth Responders for enterprise solutions to homeland security (Baldwin et al. 2010; Li et al. 2011). Reporting suspect terrorist activities can be recognized as an effective way in the phase of prevention, while the other potential capabilities could also serve as flexible and timely resources in response strategies and should be considered as part of an enterprise solution.

**Lack of Solution Ownership.** The most interesting observation that the systemigram revealed, which we do not believe was apparent in just reading the text alone, was the paradoxical tension that would exist in the realization of the SVSS via the *Maritime Governance* between the *Maritime Security Partners* and the *Small Vessel Community*. While *Maritime Governance* appears to be a construct that links these two constituencies, they have differing perspectives on the role of *Maritime Governance*. For example the *Maritime Security Partners* define their success on a *Unified Effort* via the *Maritime Governance* as executed by effective risk mitigation. However, the *Small Vessel Community* sees *Maritime Governance* as *distributed* and believes that it is their autonomous behavior that empowers them, for which to relinquish their autonomy would be to lose their independence (Podlich 2009). In addition, Major Goal A states that the *Maritime Security Partners* must *develop and leverage partnerships with* a community that does not function with the same perspective on *Maritime Governance*. What results is a paradoxical tension that will at this point render the current SVSS inoperable. This paradox demonstrates that in an enterprise, what we may believe to be the solution can also be our problem. How does the small vessel community, who may become our disruption, also serve as our solution? This is not an organization problem but an enterprise problem for which the DHS has stated that our security solutions are enterprise solutions and we must find ways to engage all levels of the enterprise. In an enterprise founded upon structure and control, paradox is designed out, or mitigated via a rigorous risk strategy. The future of enterprise resilience resides in our ability to accept paradox as a norm of enterprise behavior, and the advent of this in the systemigram was profound and encouraging. Paradox exists for a reason, and there are reasons to appreciate it. It will be our ability to govern, not control, these paradoxes that will bring new knowledge to our understanding on how to manage the emerging complexity of enterprises.

# CHAPTER 21

# TO ARRIVE WHERE WE STARTED

We shall not cease from exploration
And the end of all our exploring
Will be to arrive where we started
And know the place for the first time.

—T.S. Eliot, *Four Quartets*

This is a book about problem-solving, but with a difference. We recognized three vital characteristics, which for far too long have been overlooked or neglected in problem-solving books.

First, we identified that while solutions undoubtedly "deal with" the problems to which they relate, they also create a new wave of problems in their wake. In our complex world, this problem-generating characteristic of solutions cannot be ignored, and problem-solving itself must take care not to become problem spreading in nature. It has been widely recognized for some time that problems themselves can spread or cascade, as in the case of

*Systemic Thinking: Building Maps for Worlds of Systems*, First Edition.
John Boardman and Brian Sauser.
© 2013 John Wiley & Sons, Inc., Published 2013 by John Wiley & Sons, Inc.

electricity supply networks (e.g., the New York City blackout) or the growth of cancer in the human body (e.g., prostate cancer in adult males). But the realization of problems elsewhere caused by the creation of a solution in some particular area of interest, removed from these affected other regions, is both alarming and unsettling. The way forward that we proposed in this book gives due recognition to this phenomenon.

Second, the emergence of a class of person known as problem solver, identified by skills in problem-solving, has reduced the burden on the class known as problem owner, to the extent that the latter has effectively transferred the problem and subsequently lost ownership, and in so doing has created a false picture for the former who cannot therefore avoid endowing the solution with the problem-spreading gene. This distinction of classes, one that effectively divorces the two, must be overcome, and problem-solving in our complex world must restore the vitality of problem ownership among those who sense the problem in the first instance.

The third characteristic is something we can more easily recognize if we stand back from the first two. When a solution to a given problem also leads to a wave of new problems, then problem-solving essentially becomes problem spreading. When problem-solving attracts a new breed of people who become known as problem solvers, then responsibility for the problem is in effect transferred—from those it first affects or who sense it, with attendant diminution in problem ownership. We might say problem-solving becomes problem dispossession. So standing back leads us to conclude that the originating problem is strongly connected to a host of "accompanying apparatus," including owners, solvers, and problem-solving approaches. It is this connectedness that marks out this third characteristic that we believe has hitherto been sorely neglected and about which this book has much to say. Moreover, this book has much to offer by way of a responsive way forward.

Our way forward is what we call *systemic thinking*. It is a way of thinking that emphasizes connectedness and enables people to see the bigger picture; one in which owners, solvers, solutions, problem-solving methods, and problem descriptions are portrayed as a whole system.

Traversing this book can also be seen as a passage into Worlds of Systems. As such, the book is in three parts, which we have rightfully named Journeys. We sincerely hope that these Journeys formed a coherent whole, that when you were done you were brought to a place you were not before you started. In Journey I, we described systemic failure—an increasingly popular term among politicians *inter alia* for describing the meltdowns and near-catastrophes involving multiple stakeholders and systems—as the representation of problems that cascade. This term applies when there is evident lack of problem ownership coupled with piecemeal approaches to problem-solving and reliance on unsustainable solutions.

When confronted with a problem that appears to be without solution, we apply frameworks from our intellect to shine a light on a potential path. In Journey II, we presented a system of ideas, which helped us to form a language that better enables us to describe specific systemic failures, and in doing so form more well-rounded problem descriptions. This was our framework for enlightening a path, the Conceptagon.

In Journey III, we introduced the idea of systemic diagrams, which we call systemigrams. These are our maps to systemic problems. We provided numerous examples of how systemigrams have helped to overcome piecemeal problem-solving by drawing together owners, solvers, problem descriptions, and relevant solutions. Journey III gave a comprehensive opportunity to learn what systemigrams were, how they are created and put to effective use, and why they are an efficacious approach to complex problem-solving.

These were our journeys into Worlds of Systems and systemic thinking. Let the end begin ...

# REFERENCES

Akilesh, S. (2000). "Bioluminescence: Nature's Bright Idea." Retrieved April 16, 2013, from http://dujs.dartmouth.edu/2000S/06-Biolumen.pdf.

Baldwin, C., Q., Li, et al. (2010). "Simulating a First Responder Scenario." *Conference on Systems Engineering Research.* Hoboken, NJ.

Berlow, E. (2010). "Simplifying Complexity." Retrieved April 16, 2013, from http://www.ted.com/talks/lang/en/eric_berlow_how_complexity_leads_to_simplicity.html.

Botsman, R. (2010). *What's Mine is Yours: The Rise of Collaborative Consumption.* London, HarperBusiness.

Desira, N. (2012). "Blog 3—Weight Gain and Nutritional Guidelines." *Nicholas Desira—Systems Thinking.* Retrieved April 16, 2013, from http://nickdessystemsthinking.blogspot.com.

DHS (2010). *Quadrennial Homeland Security Review Report.* Washington, DC, Department of Homeland Security.

DHS (2011). *Small Vessel Security Implementation Plan: Report to the Public.* Washington, DC, Department of Homeland Security.

Fine, C. (1998). *Clockspeed: Winning Industry Control in the Age of Temporary Advantage*. New York, Perseus Books.

Ivory, J. (1993). The Remains of the Day, Columbia Picture Corporation.

Li, Q., B. Sauser, et al. (2011). Analyzing the Influence of Zeroth Responders on Resilience of the Maritime Port Enterprise. *IEEE Systems Conference*. Montreal, Canada.

Mansouri, M., A. Gorod, et al. (2010). "System of Systems Approach to Maritime Transportation Governance." *Transportation Research Record: Journal of the Transportation Research Board* **2166**: 66–73.

Milgram, S. (1967). "The Small World Problem." *Psychology Today* **2**: 60–67.

Nowak, M. and R. Highfield (2011). *SuperCooperators: Altruism, Evolution, and Why We Need Each Other to Succeed*, New York, Free Press.

Obama, B. (2009). Statement by the President on Preliminary Information from his Ongoing Consultation about The Detroit Incident. Kaneohe Bay Marine Base, Kaneohe, Hawaii, The White House.

"Parishes: Kempsey" (1913). A History of the County of Worcester: Volume 3. Retrieved April 16, 2013, from http://www.british-history.ac.uk/report.aspx?compid=43148.

Podlich, M. (2009). Statement of Margaret Podlich before the Coast Gaurd and Maritime Transportation Subcommittee of the Committee on Transportation and Infrastructure, United States House of Representatives. Regarding Maritime Domain Awareness, Boat Owners Association of the United States.

Priest, D. and W. Arkin. (2010). "A Hidden World, Growing Beyond Control." *The Washington Post*. Retrieved April 16, 2013, from http://projects.washingtonpost.com/top-secret-america/articles/a-hidden-world-growing-beyond-control/.

SSCI (2010, May 18). *Committee Report on the Attempted Terrorist Attack on Northwest Ailines Flight 253*. Washington, DC, Senate Select Committee on Intelligence.

Stearns, S.C. (2010, April 17). "The Impact of Evolutionary Thought on the Social Sciences." Evolution, Ecology and Behavior. Retrieved April 16, 2013, from http://videolectures.net/yaleeeb122f07_stearns_lec22/.

Strogatz, S. (2008). "Steven Strogatz on Sync." Retrieved April 16, 2013, from http://www.ted.com/talks/steven_strogatz_on_sync.html.

Texas Heart Insitute (2012). "Heart Anatomy." Retrieved April 16, 2013, from http://www.texasheart.org/HIC/Anatomy/anatomy2.cfm.

UNAIDS (2011). "Data and Analysis." Retrieved April 16, 2013, from http://www.unaids.org/en/dataanalysis/.

# INDEX

*Systemic Thinking: Building Maps for Worlds of Systems*, First Edition.
John Boardman and Brian Sauser.
© 2013 John Wiley & Sons, Inc., Published 2013 by John Wiley & Sons, Inc.

Printed and bound by CPI Group (UK) Ltd, Croydon, CR0 4YY

27/10/2024

14580272-0001